SS LANGEMARK
RECONNAISSANCE RGT

(MOT) STAFF COY

ARMOURED CAR COY

(MOT) LIGHT RECON SUPPLY
COLUMN

RECONNAISSANCE BTN

RECONNAISSANCE BTN

SS DAS REICH
FLAK BTN

(MOT) STAFF COY

1 (SELF-PROP) MEDIUM FLAK COY

2 (MOT) HEAVY FLAK BTY

3 (MOT) HEAVY FLAK BTY

4 (MOT) HEAVY FLAK BTY

(MOT) LIGHT FLAK SUPPLY COLUMN

SS DAS REICH
SIGNALS BTN

SS DAS REICH
PIONEER BTN

SS DAS REICH
PANZERJÄGER BTN

SS DAS REICH
ADMINISTRATION SERVICE

SS DAS REICH
MEDICAL SERVICE

SS DAS REICH
SUPPLY SERVICE

SS DAS REICH
MAINTENANCE BTN

SS (MOT) MILITARY
POLICE COY

DAS REICH
AT KURSK

VISUAL BATTLE GUIDE

DAS REICH
AT KURSK
12 JULY 1943

DAVID PORTER

amber
BOOKS

This edition first published in 2011

Published by
Amber Books Ltd
Bradley's Close
74–77 White Lion Street
London N1 9PF
United Kingdom
www.amberbooks.co.uk

ISBN: 978-1-907446-62-7

Project Editor: Michael Spilling
Design: Hawes Design
Picture Reseach: Terry Forshaw
Additional text: Stephen Hart

Printed in China

CONTENTS

Chapter 1
Years of Victory: Campaigns, 1939–43

In July 1943, the SS Panzergrenadier Division *Das Reich* spearheaded the titanic German armoured onslaught at Kursk, codenamed Operation 'Citadel'. Emerging back in 1934, the division's precursor units had played key roles in the early German *Blitzkrieg* successes of 1939–40; subsequently, the SS Division *Reich* achieved the bravura coup of capturing Belgrade in April 1941.

During the second half of 1941, the division spearheaded the impressive German advances through the Soviet Union; during 1942, it mounted desperate last stands against massed Soviet counter-attacks. Finally, in early 1943, now converted into a panzergrenadier division, *Das Reich* played a pivotal role in the stunning German counter-offensive at Kharkov, which set the operational context from which 'Citadel' unfolded.

OPPOSITE: Panzergrenadiers from the *Das Reich* division keep a lookout as they pass through the suburbs of Kharkov, sometime during the early spring of 1943. The soldier nearest the camera is armed with an MG42 heavy machine-gun.

Formation

The origins of what became the SS Panzergrenadier Division *Das Reich* in Operation 'Citadel' can be traced back as far as 1934, to some of the earliest Nazi Party military units.

The formation known at the time of the July 1943 German Kursk offensive as the SS Panzergrenadier Division *Das Reich* existed as a divisional-sized command from 1 April 1940 through to Germany's surrender in early May 1945. This formation began life as the *SS-Verfügungs-Division* (SS Special Readiness Division), and then became the SS-Division (motorized) *Reich*, then the SS-Panzergrenadier Division *Das Reich*, and eventually the 2nd SS Panzer Division *Das Reich*. Heavily committed to key battles such as Kharkov (February–March 1943), Kursk (July 1943) and Normandy (June–August 1944), *Das*

Members of the SS-VT Regiment take a break during the early days of the invasion of France, May 1940.

Reich displayed such an impressive combat performance that it was regarded as one of the premier *Waffen-SS* formations. The component parts of *Das Reich* predate the division's creation in April 1940, and hark right back to the earliest Nazi Party military units in 1934; and as a part of what would become the *Waffen-SS* (Combat-SS), their wider heritage can be traced back to Nazism's earliest days in the 1920s.

The *Schutzstaffel* (SS) was the vanguard political force of Adolf Hitler's National Socialist (Nazi) Party, which came to power in Germany in 1933, aiming to create a Thousand-Year Reich within a new German-dominated, Jewish-free European Order. The SS comprised those individuals most fervently committed to the Nazis'

racist, violent and expansionist ideology. The *Waffen-SS*, the SS military wing, was created in March 1940 through the amalgamation of three existing SS military institutions – the *Leibstandarte* (Hitler's bodyguard unit), the *Verfügungs-Division*, and the *Totenkopf* (Death's Head) concentration camp guard units. The *SS-Verfügungs* Division itself had been formed in early 1940 by amalgamating the existing *SS-Verfügungstruppe* (Special Readiness Troops) units. The *Waffen-SS*, *Das Reich* included, manifested the duality of being both an impressive fighting force and a brutal, fanatical, politicized movement that spearheaded the implementation of the Nazis' racist ideological goals. The volunteer recruits of *Das Reich* repeatedly demonstrated their battlefield determination and high tactical abilities in some of the most bloody and prolonged battles of World War II. But many of the division's personnel were also politically indoctrinated Nazis, who arrogantly viewed themselves as an Aryan elite with a mission to obliterate anything that stood in the way of the creation of Hitler's Reich. Imbued with an ethos of ruthlessness, *Waffen-SS* personnel regularly committed atrocities against civilians and un-armed enemy prisoners.

ORIGINS

The origins of *Das Reich* hark back to the earliest days of the Nazi regime. During the 1920s the *Sturmabteilungen* (SA) had been the Nazi Party's principal paramilitary force, but to counter-balance this in 1923 Hitler created a small personnel bodyguard, later renamed the *Stosstruppe Adolf Hitler*. Expansion of this unit led to the creation in 1925 of the *Schutzstaffel* (SS). During 1934 this bodyguard unit, now entitled the *Leibstandarte-SS Adolf Hitler,* spearheaded Hitler's bloody purge of the SA, which established the SS as the preeminent Party military force. Subsequent rapid expansion led to the creation in September 1934 of the *SS-Verfügungstruppe* (SS-VT), a militarized Party organization alongside the *Leibstandarte,* which was subordinated to neither the *Wehrmacht* (armed forces) nor the police, instead answering directly to the Nazi hierarchy. Initially, the SS-VT comprised two units, the *SS-Standarten* (Regiments) *Deutschland* and *Germania*, with a combined strength of 2250 personnel.

During the mid-1930s the militarized Nazi elements emerged as four separate organizations – the *Algemeine-SS* (General-SS), the *Leibstandarte,* the *Totenkopfverbände* concentration camp guard units, and the SS-VT. In the face of opposition from the German Army, the SS-VT initially was not only dependent on the latter for its training and equipment but was also subject to strict quotas on the personnel it could recruit. Despite these restrictions, the SS-VT raised the Battalion N, based in Nuremberg, the Pioneer Battalion *Dresden*, and a signals battalion during 1935.

During 1935–39 the *Reichsführer-SS*, Heinrich Himmler, and the Inspector of the SS-VT, SS-*Standartenführer* (Colonel) Paul Hausser, set about transforming the SS-VT into a viable party armed force to rival the German Army. Hausser established two SS officer training schools with stringent entry requirements that employed army-style training regimens. The two SS-VT regiments, as well as the specialist SS-VT battalions, got their first taste of military operations during the March 1938 German *Anschluss* (unification) with Austria, although this operation did not involve any combat. Subsequently, Hausser further expanded the SS-VT by raising the Vienna-based *SS-Standarte Der Führer* and two motorcycle-mounted infantry battalions.

By December 1938, the three regiments fielded 7762 troops between them. All three regiments and several of the battalions participated in the non-violent October 1938 German occupation of the Sudetenland, on attachment to regular army units; these actions provided SS-VT units with useful operational experience, even if again they had no opportunity to engage in combat. During the first half of 1939, the SS-VT went through considerable reorganization: the two motorcycle battalions were disbanded while an artillery regiment, an anti-aircraft machine-gun battalion and a reconnaissance battalion were raised. The SS-VT units now included most of the components found in a division, and so Hitler took the next logical step and ordered Hausser to draw up plans to amalgamate the SS-VT units into a division. However, before this could be realized, Hitler's expansionist foreign policy finally led to the commencement of World War II in Europe, in September 1939.

First Actions: Poland and France, 1939–40

Despite being trained to a high standard up to summer 1939, SS-VT units had yet to experience actual combat: this changed during 1939–40, when they participated in the early German campaigns in Poland and France.

On 1 September 1939, 17 German mobile and 39 infantry-style divisions, backed by 1900 aircraft, invaded Poland from three directions, and engaged the latter's 43 divisions/brigades and 435 aircraft. Army Group North thrust east through the Polish Corridor to link East Prussia with the Reich, and then attacked south-east towards Warsaw; meanwhile Army Group South thrust east and north-east from Silesia towards Warsaw and Lublin, while also attacking north-east from Slovakia. By 16 September these advances had encircled numerous Polish formations in several pockets. The next day, the pre-arranged Soviet assault into eastern Poland sealed the country's fate; it capitulated on 6 October. For the cost of 48,000 casualties and 674 panzers lost, the Germans had inflicted 199,800 Polish casualties and taken 588,000 prisoners. This success announced the awesome power of the fast-paced mobile warfare spearheaded by Germany's panzer divisions.

The various SS-VT units played important roles in these battles. The *Deutschland* and SS-VT Artillery Regiments, together with the SS-VT Flak MG and Reconnaissance Battalions, served in the improvised Armoured Formation *Kempf*. During 1–9 September, these units encountered bitter Polish resistance as *Kempf* advanced through Mlawa to cross the River Bug near Brok. Over the next week, these SS-VT units fought their way south-east to seal the plug in the eastern part of the German encirclement of Warsaw. Finally, on 29 September, in its last actions, the *Deutschland* Regiment helped overwhelm the

15 cm sIG 33 (Sf) auf Panzerkampfwagen I Ausf B

CREW

SPECIFICATIONS

Weight: 8.636 tonnes (8.5 tons)
Length: 4.42m (14ft 6in)
Width: 2.6m (8ft 6in)
Height: 3.35m (11ft)
Engine: 74.5kW (100hp) Maybach NL38TR 6-cylinder petrol
Speed: 40km/h (24.9mph)
Range: 140km (87 miles)
Armament: 1 x 150mm (5.9in) sIG 33 L/11 infantry gun

The Polish campaign showed that the towed 150mm (5.9in) sIG 33 guns of motorized infantry regiments had difficulties keeping up with the tanks in combat. As an interim solution, the sIG 33, complete with its carriage and wheels, was mounted in a tall, open-topped fighting compartment on the hull of the Panzer I Ausf. B. Whilst this gave the gun vastly improved mobility, it was far from perfect – the little Panzer I was severely overloaded and suffered constant breakdowns. A total of 38 vehicles were converted, serving from 1940 until the survivors were finally replaced in 1943.

resistance encountered in the Modlin-Zakrozym fortified area, north-east of Warsaw. Numerous German dispatches praised the high motivation and tactical skills of these SS-VT units. Meanwhile, the sub-units of the *Germania* Regiment had been attached piecemeal to a variety of Army Group South units, and these participated in the drive north-east through Krakow to Warsaw. These units also impressed observers with their high levels of initiative and combat motivation.

PzKfw II Ausf C

■ CREW

■ SPECIFICATIONS

Weight: 8.9 tonnes (8.76 tons)
Length: 4.81m (15ft 9in)
Width: 2.22m (7ft 3in)
Height: 1.99m (6ft 6in)
Engine: 104.4kW (140hp) Maybach HL62TR 6-cylinder petrol
Speed: 40km/h (24.9mph)
Range: 200km (124.3 miles)
Armament: 1 x 20mm (0.79in) KwK30 L/55, plus 1 x 7.92mm (0.31in) MG13 machine-gun

The Panzer II was originally intended as an interim light tank pending the introduction of the Panzer III and Panzer IV, but became one of the most important tanks of the early war years, with a total of over 1800 being produced. (A further 1250 Panzer II hulls provided the basis for a host of self-propelled guns and tank destroyers.)

SS VT Division

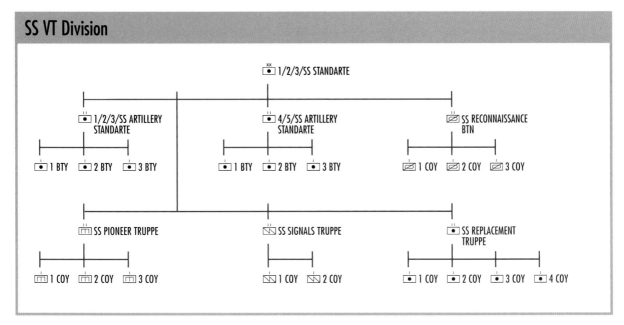

After this success Hitler began planning an invasion of north-western Europe; Germany had to clear up its rear flank before it could take on its true ideological enemy, the Soviet Union. On 10 May, the Germans initiated their invasion of France, Belgium and The Netherlands, committing a force of 131 divisions and 5600 aircraft. In the north, Army Group B thrust through the southern Netherlands and Belgium, drawing the French and British forces deployed in northern France into moving forwards into central Belgium; here the German attacks would fix them. Meanwhile, Army Group C mounted a feint against the Maginot Line defences along the Franco-German border in Alsace-Lorraine, again fixing the French forces. With the enemy pinned down to the north and south, the surprise German main effort fell on the weakly held Allied centre. Here, panzer divisions would mount a swift surprise assault through the difficult Ardennes terrain, cross the River Meuse and charge through the French rear areas towards the Channel coast, enveloping the Allied forces to the north.

By May 1940 the various SS-VT units had been amalgamated to form the semi-motorized *SS-Verfügungs-Division* (Special Readiness Division, or SS-V Division). Despite this, the formation initially participated in the Western campaign as separate units. The *Der Führer* Regiment, plus supporting elements, fought as attachments to the 207th Infantry Division, while elements of *Deutschland* served with the 254th Division; both formations came under the control of Army Group B. *Der Führer* spearheaded the German attack west through the Nijmegen-Rotterdam area, which aimed to link up with German airborne forces that had landed at Rotterdam and The Hague. After advancing 98 kilometres (61 miles) on the first day, *Der Führer* smashed through the Dutch Grebbe Line defences during 11 May, and then spearheaded the assault on Fortress Holland, the final Dutch defence line. Subsequently, the various SS-V units rejoined their parent division, and during 16–17 May overwhelmed the Dutch forces holding Walcheren island in the Scheldt estuary.

By 21 May, the panzer advance west beyond the Meuse had reached the Channel coast, creating a pocket of encircled Allied forces centred on the port of Dunkirk. The SS-V Division had meanwhile redeployed to the La Bassée Canal area of this pocket, where during 23 May its forces had to mount desperate defensive stands to repel determined French counter-attacks, which in places included armour. Next, during 26–28 May, elements of the division forced British infantry units back north from the Nieppe Forest to the Nieppe Canal. Finally, on 1 June the Germans placed the SS-V Division in reserve, where it absorbed new equipment and personnel to make good its losses. On 5 June the division was committed to the German offensive south from the River Somme, and over the ensuing 19 days it fought its way south against weak resistance through to south-western France; in these actions the division took 30,000 prisoners for the cost of just 35 casualties. By 25 June, France had surrendered.

SS Verfügungstruppe Leibstandarte Adolf Hitler, 1 April 1940

UNIT	EQUIPMENT	STRENGTH
Signals Platoon (mot)	–	1
Motorcycle Messenger Platoon	–	1
Regimental Band	–	1
1/2/3/LSSAH (mot) Infantry Battalion		
Rifle Companies (mot)	9 x LMGs	3
	2 x HMGs	
	3 x 50mm mortars	
Machine Gun Company (mot)	8 HMGs	1
	6 x 80mm mortars	
Pioneer Platoon (mot)	3 LMGs	1
Motorcycle Company	9 LMGs	1
	2 HMGs	
	3 x 50mm mortars	
Infantry Gun Company (mot)	8 x 75mm leIG	1
Panzer Abwehr Company (mot)	12 x 37mm PaK 36	1
	6 x LMGs	
Armoured Car Platoon	4 x 20mm	1
	8 x LMGs	
1/2/3/SS Standarte Deutschland, Germania, Der Führer		
1/2/3/SS Artillery Standarte	–	3 batteries
4/5/SS Artillery Standarte	–	3 batteries
SS Reconnaissance Battalion	–	3 companies
SS Pioneer Truppe	–	3 companies
SS Signals Truppe	–	2 companies
SS Replacement Truppe	–	4 companies

Panzerabwehrkanone 36 (PaK 36)

■ CREW

■ SPECIFICATIONS

Weight in action: 0.328 tonnes
(0.322 tons)
Calibre: 37mm (1.45in)
Muzzle velocity: 760m/s (2495 ft/sec)
(AP)
Range (HE): 7000m (7655 yards)
Ammunition: AP, HE

The PaK 36's origins date back to 1924, when Rheinmetall designed a 37mm (1.45in) horse-drawn anti-tank gun. This entered service with the *Reichswehr* in 1928 as the 3.7cm PaK L/45 and was redesigned in the early 1930s to make it suitable for towing by motor vehicles. The new gun was designated PaK 36 and began to replace the earlier weapon in 1934. It was widely exported and the design formed the basis for the development many other nations' anti-tank guns in the immediate pre-war period. Whilst it was effective against the AFVs of the 1930s, it was outclassed by British and French heavy tanks in 1940. The gun's uselessness against the T-34 in the opening stages of Operation *Barbarossa* led to it being dubbed 'the doorknocker' and it was rapidly replaced by larger-calibre anti-tank guns.

SdKfz 231 (6 Rad)

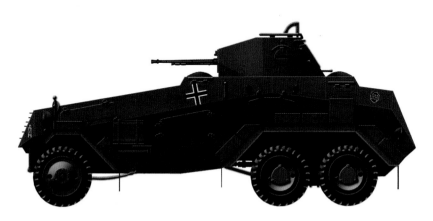

■ CREW

■ SPECIFICATIONS

Weight: 5.791 tonnes (5.7 tons)
Length: 5.61m (18ft 4in)
Width: 2.24m (7ft 4in)
Height: 1.85m (6ft 1in)
Engine: 48kW (65hp) Daimler-Benz M09
6-cylinder petrol
Speed: 65km/h (40mph)
Range: 250km (150 miles)
Armament: 1 x 20mm (0.79in) KwK 30
L/55, plus 1 x 7.92mm (0.31in) MG34
machine-gun

The SdKfz 231 design was based on a Daimler-Benz 6x4 truck chassis modified to allow steering from either end and fitted with an armoured hull and turret. Initial trials demonstrated the need for a stronger front axle and improved radiator, and the first production vehicles entered service in 1932. Production continued until 1935, by which time about 1000 had been completed. These armoured cars had poor cross-country performance as a result of their 6x4 configuration and were under-powered, but they were as good as most of their foreign counterparts of 1939–40. From 1941, the surviving vehicles were gradually relegated to training and anti-partisan duties.

Reich SS (mot) Division, 1940–41

After its promising successes during the May 1940 German invasion of the West, the semi-motorized SS-V Division was converted into a fully fledged motorized division, enabling it to spearhead the future German *Blitzkrieg* campaigns of 1941.

Between late 1940 and early 1941, the *SS-Verfügungs* Division, still under Paul Hausser's command, was stationed in the Vesoul area of central-eastern France. While deployed here, the division gave up significant cadres to other SS formations, and also absorbed new personnel and equipment as part of a major reorganization. During this process the division was transformed from a semi-motorized formation into a fully-fledged infantry division (motorized). In late November the formation was renamed the SS-Division *Deutschland*, but on 21 December it was renamed SS-Division (motorized) *Reich*. The

division then lost the *Germania* regiment, which went to form a new SS-Division, originally named *Germania* but subsequently re-titled *Wiking*. In its place *Reich* received the motorized 11th SS Infantry Regiment, which formerly had been 11th SS-*Totenkopf* Regiment.

Thus the *Reich* still had three regiments, each of three battalions, as its main combat strength, although each regiment now fielded 16 companies instead of 13. The division also lost an NCO cadre dispatched to the *Leibstandarte* to aid its expansion into a reinforced infantry brigade (motorized). The final newly raised unit added to

ORBAT: *Reich* SS (mot) Division (December 1940)

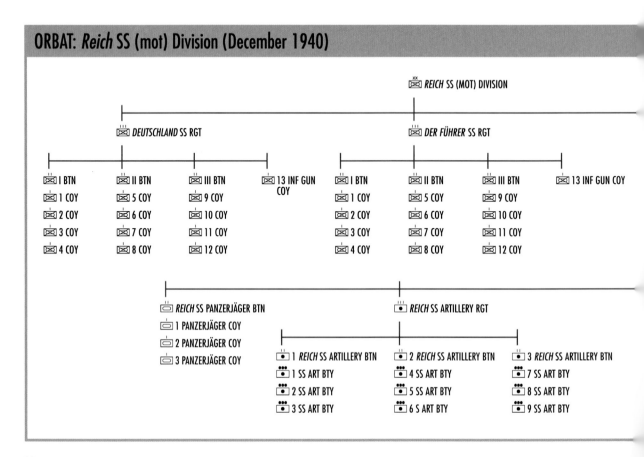

the division was an assault gun battery that fielded six StuG III vehicles.

The division, meanwhile, continued to field its existing ancillary units, all now entitled *Reich*. Some of these were expanded; the artillery regiment, for example, raised a fourth battalion, while the motorcycle reconnaissance battalion grew from three to four companies. On completion of this reorganization, the division now also deployed a three-company anti-tank battalion, a three-battery flak battalion, a three-company pioneer battalion, a two-company signals battalion, and the usual supporting elements, many expanded in size (which included military police, supply, vehicle repair, medical, veterinary, replacement, butcher, and field bakery detachments). The division's war establishment was now about 19,000. This made it more powerful than the standard German army infantry division (motorized), as the latter generally only had two infantry (motorized) regiments and a three-battalion artillery regiment. By March 1941

the Germans judged that the transformed *Reich* Division was adequately combat-ready and thus redeployed it Romania.

The division now joined the dozen or so German infantry divisions (motorized) that were intended to serve alongside the panzer divisions as the spearhead of what the victims of Germany's actions later dubbed *Blitzkrieg*, or Lightning War. In this operational method, the Germans concentrated their elite panzer and motorized divisions at a few key points of the battle-space; with the aid of intimate air support, these mobile forces quickly broke through the enemy's frontier defences. Subsequently, these forces developed audacious fast-paced strategic penetrations designed to envelop and encircle large chunks of the enemy's order of battle. By the time the following non-motorized infantry divisions had arrived to complete the defeat of these demoralized pockets of enemy forces, the mobile spearheads had continued to dash deep into the

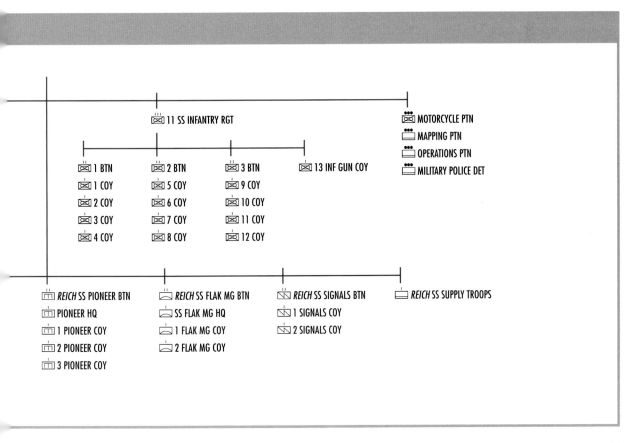

enemy's rear areas, collapsing the latter's will and cohesion to deliver a quick and decisive defeat. The Germans named this approach 'Cauldron Battle', after the cauldrons (or pockets) created by these high-tempo encirclement operations.

As part of this approach, the German motorized infantry divisions typically served within a panzer corps alongside armoured divisions. More rarely, they also served within the few motorized corps that existed, alongside other motorized divisions. Within a panzer corps, motorized divisions generally undertook the role of screening and reconnaissance to provide flank protection, thus enabling the spearhead armour to be concentrated at the key point (or *Schwerpunkt*). In other tactical situations, the motorized regiments of motorized divisions might be attached to a nearby panzer division – to provide, for example, additional infantry strength when operating in built-up areas, which put a premium on infantry providing close co-operation with the tanks. Finally, motorized divisions might also be used as the reserve element within a panzer corps, following up the spearhead forces to mop up bypassed or surrounded enemy units, to hold key pieces of ground, or to protect the lines of communication.

Balkan *Blitzkrieg*, April 1941

During spring 1941, Hitler's preparations for the future invasion of the Soviet Union were disrupted by the need to re-stabilize the Balkans through an invasion in which the *Reich* Division played a stunning role.

During winter 1940–41, Germany needed to ensure a peaceful Balkans, which would not disrupt the lengthy preparations required for Operation *Barbarossa*, its planned invasion of the Soviet Union, slated for May 1941. On 28 October 1940, however, the Italians – wishing to emulate Germany's recent martial glory – invaded north-western Greece from Albania. By 14 November, however, Greek counter-attacks had forced the Italians back onto Albanian soil, and Hitler now feared that Greece's British ally would intervene militarily, jeopardizing the prospects of *Barbarossa*.

In early 1941, Hitler decided to resolve this issue by mounting a German invasion of Greece, launched from the soil of her Axis partners, Romania and Bulgaria. Next, Greece accepted an offer of two British divisions on its soil, while on 27 March a coup in Yugoslavia installed a pro-Allied regime; these events prompted Hitler to order a hastily improvised invasion of both Yugoslavia and Greece commencing on 6 April. Involving 10 panzer, four motorized, and 18 infantry-type divisions, the invasion began with ferocious aerial bombing strikes on the Yugoslavian capital, Belgrade, despite it being declared an 'open city' free of military units. Although the Yugoslav Army was numerically strong, having 28 divisions, it possessed no armoured units, had mostly old equipment, suffered from poor formation cohesion thanks to fierce ethnic tensions, and enjoyed little air support.

During 6–11 April, various Axis assaults invaded Yugoslavia from three sides. During 10–11 April, along the north-western flank, Italian forces in Istria thrust south-east into western Croatia while across the northern front German forces attacked south from southern Austria towards Zagreb in central Croatia. On 11 April, on the north-eastern front, XLVI Panzer Corps struck west from southern Hungary towards Zagreb.

LONG MARCH

Along the eastern front on 11 April, meanwhile, XLI Panzer Corps, which included the SS-Division (motorized) *Reich*, attacked from western Romania, thrusting south-west towards Belgrade, located just 92 kilometres (57 miles) away. The *Reich* Division had commenced its long road journey from France only on 28 March and it reached the Temesvar region of Romania on 4 April. Along the way, several altercations arose with redeploying German Army units on the matter of who had priority for movement, displaying the tensions that existed between the *Wehrmacht* and *Waffen-SS*. Meanwhile, on 7 April, along Yugoslavia's south-eastern borders, Panzer Group 1 had attacked from western Bulgaria, striking north-west towards Nis and Belgrade. The previous day the German Twelfth Army had attacked south-eastern Yugoslavia from south-western Bulgaria, thrusting south-west to link up with the Italian advance north-east, mounted from eastern Albania; at the same time, Twelfth Army also struck south to invade northern Greece.

The Germans made good initial progress on all fronts, aided in the north-west by the defection of Croat units within the Yugoslav Army. By late on the 10th the Germans had captured Zagreb, and immediately Croat Nationalists announced the creation of a pro-German Croatian Fascist state. By 12 April, meanwhile, Panzer Group 1 had advanced northwest to secure the high ground

Hungarian officers look on as a motorcycle combination from the *Das Reich* division speeds past during a formal parade through Budapest, just before the invasion of Yugoslavia in April 1941.

south-west of Belgrade. On the previous day XLI Corps had initiated its thrust south-west from western Romania towards Belgrade. To motivate his forces to develop as rapid an advance as possible, the corps commander had stated that the first formation to reach the Alibunar-Belgrade highway would be allowed to spearhead the corps' assault on the city: divisional commander Paul Hausser was determined that his forces would win this privilege.

During the morning of 11 April, however, as the main combat elements of the Division *Reich* advanced south-west towards Jermenovici and Margita, they became bogged down in the marshy terrain that surrounded the Viasecki Canal. Luckily for Hausser, however, the division's motorcycle battalion managed to maintain its forward momentum by driving along the tops of dikes to reach the main Alibunar-Belgrade road early that afternoon; within four hours the rest of *Reich* had secured the town of Alibunar itself. Having reached the highway, the lead elements of *Reich* advanced swiftly down the road through Pancevo to approach the northern bank of the Danube opposite the

Yugoslav capital. Here, the division was ordered to halt.

On the morning of 12 April 1941, *SS-Hauptsturmführer* (Captain) Fritz Klingenberg led a 10-man section from his 2nd Company, part of the Motorcycle Recce Battalion *Reich,* on a reconnaissance patrol along the Danube's northern banks. Klingenberg was anxious to reconnoitre into the open city, but the swollen river and lack of any usable bridges prevented a crossing, and his company had no bridging equipment or rafts. Through good fortune, the patrol discovered a motor launch on the river's northern bank; Klingenberg and eight men crossed the Danube in this vessel. When the group had reached the other side, Klingenberg sent two men back to obtain reinforcements and then he proceeded with the remaining six men into central Belgrade; after it had

ORBAT: *Das Reich* SS (mot) Division (April 1941)

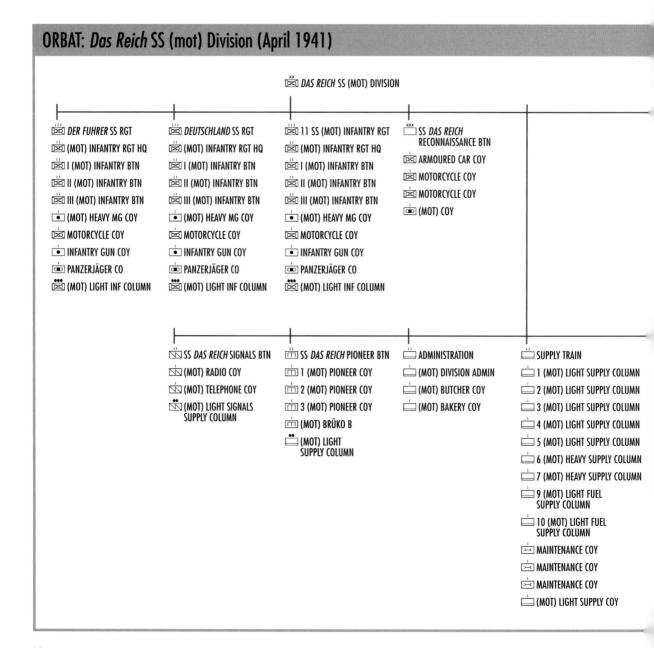

been declared an open city, virtually all the defending Yugoslav forces had moved out from the city. Soon after entering Belgrade, however, Klingenberg encountered a small group of Yugoslavian soldiers and without firing took their surrender. Later on, a group of military vehicles approached Klingenberg's men, and after a short fire-fight, the Germans captured the vehicles. The now motorized assault detachment then headed towards the Yugoslav War Ministry, but when they arrived they found the building abandoned.

Since there was no enemy military command left in Belgrade, Klingenberg proceeded to the German embassy, which had remained open despite the hostilities. At 17:00 the Germans unfurled a large swastika flag and raised it over the embassy to proclaim the capture of the city. Klingenberg persuaded the German Military

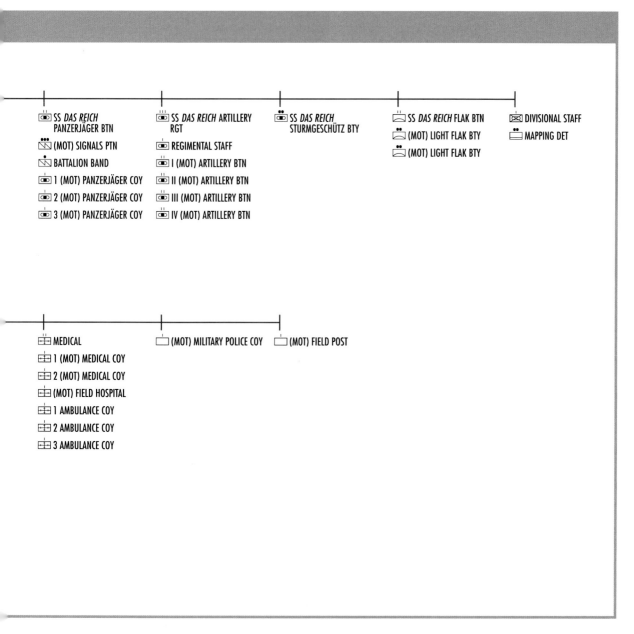

Attaché to summon the Mayor of Belgrade to the embassy. At their meeting, Klingenberg claimed that a large German force was about to attack the city on the backs of yet another ferocious aerial bombing strike – unless the Mayor surrendered the city. At 18:50, the Mayor reluctantly acceded to Klingenberg's ultimatum. It was not until later that evening that a sizeable German force – the spearhead of 11th Panzer Division from Panzer Group 1 – arrived from the south-east to secure the city. For capturing Belgrade in such dashing fashion, Klingenberg became an instant propaganda hero and was subsequently awarded the highly prized Knight's Cross, while the *Reich* division gained a reputation for breathtaking audacity.

SWIFT CONQUEST

By 13 April, in the face of the Axis attacks launched from virtually all sides, the morale and combat cohesion of many units within the Yugoslav Army had all but disintegrated. The German forces now advanced at alarming speed throughout Yugoslavia, capturing Sarajevo, the capital of Bosnia-Herzegovina, on the 15th. Realizing that continued resistance was futile, the

Yugoslavs now requested an armistice, which was signed on 17 April. The Germans had conquered Yugoslavia quickly and with amazingly light casualties. Yet this victory was less complete than it appeared: the Germans scarcely controlled many of the mountainous areas, within which a ferocious anti-Axis partisan struggle would subsequently rage until May 1945.

In the wake of the conquest of Yugoslavia, German forces poured south to reinforce the earlier German invasion of north-eastern Greece. By 23 April, the remaining Greek forces still holding out on the mainland had surrendered, although resistance continued on Crete until 1 June. Hitler had decisively ensured that Balkan politics could no longer disrupt his planned invasion of the Soviet Union, while the *Reich* Division had demonstrated the immense military potential of fanatical and risk-embracing SS units. The dramatic seizure of Belgrade represented a fitting finale for the division's involvement in the Balkans. Subsequently *Reich* returned first to Romania and then redeployed to Austria, where it undertook further unit training prior to its next mission, Hitler's invasion of the East.

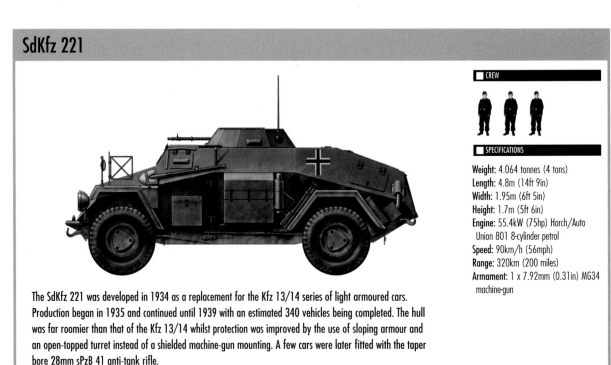

SdKfz 221

CREW

SPECIFICATIONS

Weight: 4.064 tonnes (4 tons)
Length: 4.8m (14ft 9in)
Width: 1.95m (6ft 5in)
Height: 1.7m (5ft 6in)
Engine: 55.4kW (75hp) Horch/Auto Union 801 8-cylinder petrol
Speed: 90km/h (56mph)
Range: 320km (200 miles)
Armament: 1 x 7.92mm (0.31in) MG34 machine-gun

The SdKfz 221 was developed in 1934 as a replacement for the Kfz 13/14 series of light armoured cars. Production began in 1935 and continued until 1939 with an estimated 340 vehicles being completed. The hull was far roomier than that of the Kfz 13/14 whilst protection was improved by the use of sloping armour and an open-topped turret instead of a shielded machine-gun mounting. A few cars were later fitted with the taper bore 28mm sPzB 41 anti-tank rifle.

Barbarossa: Invasion of Soviet Russia, June 1941

In June 1941 the Germans initiated their titanic ideological war of annihilation against the Communist Soviet Union, a crusade fittingly spearheaded by the SS-Division *Reich.*

On 22 June 1941 Operation *Barbarossa,* the Axis invasion of the Soviet Union, commenced. The Nazis perceived the Communist Soviet Union, populated by Slavic sub-humans, as the chief threat to the Reich, and wanted *Lebensraum* (living space) in the east to obtain economic self-sufficiency. By 22 June, the Germans had amassed on the Soviet border 139 divisions – with 3.05 million troops and 4204 AFVs – together with 22 Axis divisions. Army Groups North, Centre and South controlled these forces, with four panzer groups – amounting to 17 panzer and 12 motorized divisions – spearheading the invasion: two such groups (the 2nd and 3rd) led Army Group Centre's assault, with the SS-Division (motorized) *Reich,* still commanded by Paul Hausser, serving in General Heinz Guderian's 2nd Panzer Group.

Against this onslaught, the 190 Red Army divisions deployed in European Russia fielded 2.6 million troops and 11,500 tanks, although just 1400 of these were modern. The Axis invasion plan envisaged that the panzer groups, through audacious high-tempo advances, would encircle large groupings of Soviet forces. Once the panzers had formed these encirclements, some would remain to seal the pocket to prevent the surrounded forces from withdrawing deeper into the Soviet interior, while the remainder would continue to advance east; in the meantime the infantry divisions would march forwards to mop-up the pockets.

During the first 10 days of the invasion the four panzer groups, after rapidly breaching the border defences, charged eastwards against modest resistance. On 28 June, Army Group Centre, which

PzKpfw III Ausf F

■ CREW

■ SPECIFICATIONS

Weight: 19.5 tonnes (19.2 tons)
Length: 5.38m (17ft 8in)
Width: 2.91m (9ft 7in)
Height: 2.45m (8ft)
Engine: 223.7kW (300hp) Maybach HL
  120 TRM V12 petrol
Speed: 40km/h (24.9mph)
Range: 165km (102.5 miles)
Armament: 1 x 50mm (1.97in) KwK38
  L/42, plus 2 x 7.92mm (0.31in) MG
  34 machine-guns

In January 1934, specifications were issued for a medium tank with a maximum weight of 24 tonnes (23.6 tons) and a top speed of 35km/h (21.75mph). It was intended as the main battle tank of the Panzer divisions, with the primary role of engaging and destroying opposing AFVs. Daimler-Benz, Krupp, MAN and Rheinmetall all produced prototypes, which underwent trials during 1936 and 1937. The Daimler-Benz design was selected and the first production model, the Ausf A, came off the assembly lines in May 1937. However, mass production of the type only really began in 1939 with the improved Ausf F.

had initially held *Reich* in reserve, now comitted it to force a river crossing near Dukora and capture Starzyca, securing XLVI Panzer Corps' northern flank. By 1 July, the 2nd and 3rd Panzer Groups of had advanced 401km (249 miles) east to cross the River Beresina, having successfully encircled Soviet forces near Bialystok and Minsk. As Guderian's forces advanced through the Minsk region, *Reich* fought a series of fierce actions along the northern fringes of the Pripet Marshes. Subsequently, during 1–19 July, the 2nd and 3rd Panzer Groups successfully advanced a further 290km (180 miles) towards Moscow, to all but encircle another 300,000 Soviet troops at Smolensk. *Reich*, meanwhile, had fought itself forward to cover XLVI Corps' northern flank before spearheading the assaults on Yelnya launched during 19–21 July.

At this point, after a much-needed logistical pause, Army Group Centre did not resume its drive on Moscow. Instead Hitler ordered the 2nd Panzer Group to deal with the threat posed to Army Group Centre's southern flank by the still-cohesive Soviet forces in the Ukraine. But before Guderian's forces could redeploy, they had to fight off a series of fierce Soviet counter-attacks around Yelnya between 19 July and 8 August. *Reich* was heavily engaged in a series of desperate defensive stands against these counter-attacks, during which it sustained 1663 casualties. The division then briefly went into reserve to allow it to absorb personnel replacements, and repair the hundreds of its vehicles that were now unserviceable due to the wear of driving along hundreds of miles of un-metalled roads and tracks. Next, between 22 August and 15 September, Guderian's forces, including *Reich* as the western flank force in XIV (Motorized) Corps, advanced 396km (246 miles) south to Lokhvitsa, deep into the rear of the Soviet forces defending Kiev; here they linked up with the eastward advance of the 1st Panzer Group from Army Group South to encircle 500,000 Soviet troops.

With this threat neutralized, Hitler ordered that the attack on Moscow be renewed, and so the 2nd Panzer Group, including *Reich*, redeployed during the second half of September from the Ukraine back to its original late-July positions near Smolensk; the *Reich* then went into reserve for a few days to absorb replacements. From 30 September, Army Group Centre – now spearheaded by three panzer groups with 1350 tanks – commenced Operation 'Typhoon'. By 11 October, the German armour had encircled 660,000 Soviet troops in the Bryansk and Vyazma pockets, decimating the Soviet defences covering Moscow. Committed on 4 November as part of XLVI Corps, *Reich* played a key role in these operations, spearheading the capture of Gshatsk on the 9th. Just as the panzers began to race towards the enemy capital, however, torrential autumn rains turned the ground into a quagmire, thus bogging down their advance. Despite this, the German forces continued to grind their way eastwards, fighting through appalling mud against the fanatical

Soldiers from the Reich division cross a river during the invasion of Soviet Russia, August 1941. They are dressed in their green and brown spring pattern camouflage smocks and helmet covers.

Operation *Barbarossa*

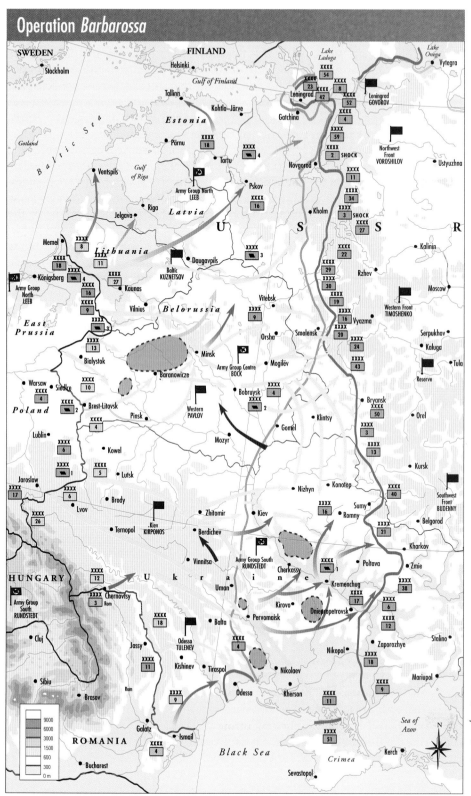

22 JUNE – EARLY OCTOBER 1941
The German plan involved three army groups (North, South and Centre), with the bulk of the forces concentrated in Army Groups North and Centre. Army Group Centre, which contained around half the German armour, was to shatter Soviet forces in Belorussia before turning to assist Army Group North in the drive on Leningrad. Army Group South, meanwhile, was to deal with Soviet forces in the Ukraine. At 03:05 on 22 June, Army Group North began the drive to Leningrad. By the evening of the first day, the leading Panzers were 60km (37 miles) into Lithuania. By the end of the second day, only the wrecks of 140 Soviet tanks lay between the Panzer divisions and Pskov. But the *Panzergruppe*'s infantry could not keep up. The terrain encountered on the Soviet side of the border was so marshy and impenetrable that even the motorized infantry were reduced to the pace of the marching columns.

Operation *Barbarossa*
22 June – early October 1941

→	German attack
XXXX 6	Soviet positions, 22 June
⬭	Soviet units encircled
←	Soviet counter-attacks
～	German front line, end of August
～	German front line, early October
XXXX 6	Soviet positions, early October

resistance offered by Soviet troops recently transferred from Siberia; *Reich,* now commanded by SS-*Brigadeführer* (Brigadier) Wilhelm Bittrich, encountered fierce Soviet counter-attacks around Mozhaisk and Pushkin, which were repelled only after the bitterest hand-to-hand fighting. Eventually, the exhausted German forces had temporarily to halt 'Typhoon' on 25 October, so that they could reorganize themselves in preparation for a renewed thrust on Moscow; the Soviets meanwhile hastily improvised new defences in front of their capital.

On 15 November, after the first winter frosts had hardened the waterlogged ground, the Germans recommenced 'Typhoon', advancing east to the north and south of Moscow, aiming to encircle the city. By then, however, Army Group Centre's 11 panzer divisions fielded just 900 tanks, a mere 35 per cent of their strength on 22 June. From the outset, German progress was painfully slow, as determined Soviet resistance, serious fuel problems and the appalling cold degraded their combat power. It took XLVI Corps, with *Reich* as its southern spearhead, some 12 days to close in on and capture

the key town of Istra. Nevertheless, by 1 December the most advanced spearhead of the northern German thrust – the Division *Reich* – had painfully fought itself forward to capture the Moscow suburb of Lenino, just 11 miles (18km) from the Kremlin. But the exhausted Germans could not continue their offensive, since no panzer division now fielded even 10 serviceable tanks.

Worse still, on 6 December the Soviets unleashed a strategic counter-offensive that aimed to destroy Army Group Centre. Over the next three weeks, repeated Soviet attacks compelled *Reich* to undertake a bitter fighting withdrawal from the River Istra back to the River Rusa. The failure of *Barbarossa* locked the Germans into an attritional war that they stood little chance of winning; during 1941 the Axis had inflicted 4.3 million casualties on the Red Army, yet they now faced an enemy more powerful than it had been back in June! Worse still, the Germans had suffered 780,000 personnel casualties and lost 3370 AFVs. This blood-letting was even starker within the *Reich* Division, which had manpower losses of a staggering 60 per cent.

Panzerkampfwagen KV-IA 753(r)

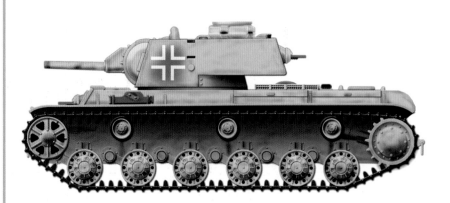

CREW

SPECIFICATIONS

Weight: 43 tonnes (42.3 tons)
Length: 6.68m (21ft 11in)
Width: 3.32m (10ft 11in)
Height: 2.71m (8ft 11in)
Engine: 447.42kW (600hp) V-2K diesel
Speed: 35km/h (21.75mph)
Range: 335km (208 miles)
Armament: 1 x 76.2mm (3in) L-11, plus
2 x 7.62mm (0.3in) DT machine-guns

The KV-1 and T-34 made an enormous impression on German forces during the early stages of Operation *Barbarossa*. Any vehicles that were captured in operational condition were taken into service, after being painted with numerous over-sized national markings in an attempt to minimize the risk of friendly fire. This KV-1 Model 1940, operated by *Reich* in the winter of 1941/42, has been fitted with a cupola from a Panzer III or Panzer IV to improve the commander's field of vision when operating 'closed down'.

Prelude to Kursk: Kharkov, March 1943

As dangerous Soviet offensives threatened to overwhelm the southern German front during February–March 1943, *Das Reich* played a key role in the counter-attacks that stabilized the situation.

The start of 1942 found the Division *Reich* locked in bitter defensive actions against the continuing Soviet counter-offensive around Moscow. On 16 January, the Germans ordered the formation to retreat back to a new defensive line west of Gshatsk, abandoning the territory obtained at such heavy cost the preceding November. Towards the end of the month *Reich* participated in a savage counter-attack that encircled the Soviet forces that had broken through to Rzhev. Over the ensuing fortnight, amid exceedingly cold weather, the division made a series of vicious defensive stands to fight off the repeated Soviet relief and break-out attempts.

By 28 February these fierce engagements had cost the division over 4000 casualties, reducing it to a mere battle group; the *Der Führer* Regiment was so depleted it had an effective strength of just one company! With its personnel pushed to the limits of human endurance, the German command was compelled to put Reich into reserve so that it could recover and absorb 3000 new replacements. By this time, as SS-*Obersturmführer* (Lieutenant) Hans Werner Woltersdorf recalled, his platoon comrades were so filthy from the appalling frontline conditions that they were all infested with lice.

Returning to the front in March, the battle group *Reich* was again heavily committed for the next four weeks in a series of desperate defensive last stands along the River Volga; by this time the division had a new commander, SS-*Gruppenführer* (Major-General) Georg Keppler. Subsequently, during July–August 1942, the division was sent to France to refit and reorganize itself as a mechanized (Panzergrenadier) division; during this process it was re-titled *Das Reich*. The new units earmarked for the mechanized division, such as SS-Panzer Regiment 2, had already begun to be raised from April at training grounds located in Germany. In the middle of this rebuilding process, *Das Reich* participated in the German occupation of Vichy-controlled southern France. Then, on 9 November, the formation was re-titled *SS-Panzergrenadier-Division Das Reich*. It now contained the Panzergrenadier regiments *Deutschland* and *Der Führer*, as well as a new rifle regiment titled *Langemarck*, which was formed from the existing motorcycle battalion and *Totenkopf* units. The division also contained the newly raised SS-Panzer Regiment 2 and an assault gun battalion. In addition, *Das Reich* continued to field its existing support units. Once fully equipped and manned, the new formation spent the next six weeks undertaking intensive individual and collective training in its new role as a mechanized division.

Das Reich panzergrenadiers climb aboard a StuG assault gun during the Kharkov campaign, March 1943.

While *Reich* refitted in France, the bitter struggle on the Eastern Front raged unabated. On 28 June 1942 the Germans initiated an offensive that aimed to secure the oil-rich northern Caucasus. Between September and mid-November, the German Sixth Army was sucked into house-by-house combat in the streets of the key communications hub and psychological prize of Stalingrad. By 23 November, however, a Soviet counter-offensive had surrounded 220,000 Axis troops in the city. While the *Luftwaffe* attempted to supply the surrounded army, the attacks of Erich von Manstein's newly constituted Army Group Don failed to re-establish a ground corridor to Sixth Army during 12–24 December. By then a powerful Soviet offensive that aimed to advance south-west to Rostov to encircle Army Group A in the northern Caucasus had smashed Manstein's northern sector. These advances sealed the fate of Sixth Army at Stalingrad; during 1–3 February 1943, the remaining 91,000 troops capitulated.

Between late December 1942 and mid-January 1943, Soviet armour pushed rapidly south-west towards Rostov. By then Soviets had resumed their attacks against Manstein's northern flank, aiming to recapture Belgorod, Kursk and Kharkov, as well as to push further south-west to the Sea of Azov to encircle the bulk German forces in the Mius River area. By 8 February 1943, the Soviets had recaptured Kursk and had begun to encircle Kharkov, the Ukrainian capital, in what would be the third battle for this city. The defence of Kharkov, which Hitler now ordered be held to the last man, fell to SS-*Obergruppenführer* (General) Paul Hausser's recently formed SS-Panzer Corps. This command controlled the elite SS-Panzergrenadier Divisions *Das Reich* and *Leibstandarte*, which had only just arrived in the East after completing their conversion process to mechanized divisions in Nazi-occupied France.

Hitler's order to hold Kharkov was nonsensical: If Hausser obeyed, *Das Reich* and *Leibstandarte* would be annihilated, and this loss would do nothing to re-stabilize the Axis position in the south. Hausser, however, bravely disobeyed his *Führer* to save his command from a destruction that served no useful strategic purpose. By 15 February, the Soviets had all but encircled *Das Reich* and *Leibstandarte* in Kharkov and so, that night,

Hausser on his own initiative commenced a fighting withdrawal from the city back to new positions on the River Uday. On hearing of this disobedience, Hitler flew into a rage, but eventually even he recognized the logic of his subordinate's actions. After all, the determined resistance offered by *Das Reich* and *Leibstandarte* at Kharkov had bought sufficient time to enable other German units to re-establish a defensive line west of the city. *Das Reich* now spent a short period in reserve to rest and absorb new personnel reinforcements. During 11–16 February, meanwhile, Soviet armour had advanced west beyond Kursk, south-west towards the River Dnepr near Zaporozhye, and south towards Mariupol on the Sea of Azov coast. On the 17th, Soviet armour reached Pavlograd and Krasnoarmeyskoye, the latter just 158km (98 miles) short of the coast, a development that portended the collapse of the entire Axis southern front.

AVOIDING DEFEAT

Between mid-February and mid-March 1943, however, Manstein's forces skilfully conducted defensive-offensive operations against the over-extended Soviet spearheads, and these delivered the miracle required to avoid decisive defeat. On 18 February, Manstein unleashed his counter-offensive with XL Panzer Corps counter-attacking at Krasnoarmeyskoye. Next, during 19–23 February, the SS-Panzer Corps, now reinforced by the recently arrived SS-Panzergrenadier Division *Totenkopf*, thrust south from Krasnograd towards Pavlograd to link up with the north-easterly advance of XLVIII Panzer Corps from south of Pavlograd, thus encircling sizable Soviet forces. Through the combination of audacity and determination, all-arms battle groups from *Das Reich* played a crucial part in these actions by securing a key bridgehead over the Woltschia River near Pavlograd on the 22nd. Next, between 24 February

OPPOSITE: The German winter counter-offensive achieved both strategic and tactical surprise. On the strategic level, this was due to the fact that all the operational planning was carried out at Manstein's HQ, which gave immunity from the Soviet 'Lucy' spy ring that had infiltrated OKH. At the tactical level, on the other hand, the Red Army had become overconfident, believing that the *Wehrmacht's* losses at Stalingrad had left it incapable of making any significant counter-attacks.

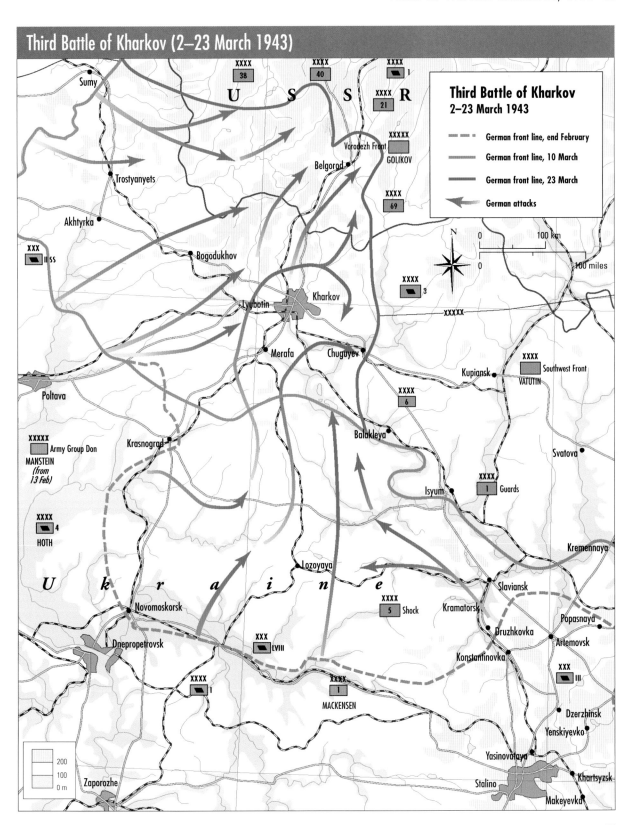

Third Battle of Kharkov (2–23 March 1943)

Third Battle of Kharkov
2–23 March 1943

- – – – German front line, end February
- ——— German front line, 10 March
- ——— German front line, 23 March
- ← German attacks

0 _____ 100 km
0 _____ 100 miles

N

Sumy

U   S   S   R

XXXX
38

XXXX
40

XXXX
21

XXXX
1

Trostyanyets

XXXXX
Voronezh Front
GOLIKOV

Belgorod

Akhtyrka

XXXX
69

XXX
II SS

Bogodukhov

XXXX
3

Lyubotin   Kharkov

Merafa   Chuguyev

Kupiansk   Southwest Front
VATUTIN

XXXX
6

Poltava

XXXXX
Army Group Don
MANSTEIN
(from
13 Feb)

Krasnograd

Balakleya

Svatova

XXXX
1   Guards

Isyum

XXXX
4
HOTH

Lozovaya

Kremennaya

U   k   r   a   i   n   e

Slaviansk

Novomoskorsk

XXXX
5   Shock

Kramatorsk

Popasnaya

Druzhkovka   Artemovsk

Dnepropetrovsk

XXX
LVIII

Konstantinovka

XXXX
1

XXXX
1
MACKENSEN

XXX
III

Dzerzhinsk

Yenskiyevko

Yasinovataya

200
100
0 m

Zaporozhe

Stalino   Khartsyzsk

Makeyevka

Sturmgeschütz III (StuG III) Ausf D

CREW

SPECIFICATIONS

Weight: 22 tonnes (21.65 tons)
Length: 5.4m (17ft 8.5in)
Width: 2.95m (9ft 8in)
Height: 1.96m (6ft 5in)
Engine: 197.6kW (265hp) Maybach
  HL120 TRM 12-cylinder petrol
Speed: 40km/h (24.9mph)
Range: 165km (102 miles)
Armament: 1 x 75mm (2.95in) StuK37
  L/24 gun, plus 1 x 7.92mm (0.31in)
  MG34 machine-gun

Prototypes of the StuG III appeared in 1937 – the first of 9500 vehicles that would be produced by March 1945. The low, well-armoured design was a great success – initially all StuGs were armed with the low-velocity 75mm (2.95in) L/24 and were employed in the direct-fire infantry support role, but later variants fitted with longer barrelled 75mm guns (L/43 and L/48) were formidable tank destroyers, accounting for a total of 20,000 Allied AFVs by the spring of 1944.

and 3 March, Manstein's forces, with *Das Reich* still in a spearhead role, advanced north towards the Donets River on a broad front, trapping Soviet forces west of the River. Manstein's daring counter-offensive had excised the dangerous Soviet salient beyond the Donets and inflicted 70,000 casualties on the enemy.

Subsequently, from 4 March, and despite having suffered terrible casualties – *Das Reich* had just 13 operational tanks – Hausser's divisions attacked north-east towards Kharkov and Belgorod before the spring thaw turned the frozen ground into a quagmire. By 8 March Hausser's spearheads had fought their way into Kharkov's suburbs, initiating the fourth battle for this city. Manstein now ordered Hausser to encircle Kharkov and only then launch probing attacks into the city to test the Soviet defences. Hausser's reconnaissance patrols led him to believe that a sudden surprise assault might work if he struck immediately. Consequently on 11 March elements of the *Das Reich* and *Leibstandarte* Divisions fought their way into the centre of Kharkov in extremely bitter and costly street fighting, even before the rest of Hausser's Corps had completed the encirclement of the city. Two all-arms battle groups built around the *Der Führer* and

Deutschland Regiments spearheaded the attacks mounted by *Das Reich,* which reached Kharkov's central railway-station by noon on 12 March. Hausser now halted his attack on the city-centre and instead redeployed his forces to complete the city's encirclement. Only after this had been achieved on 15 March did *Das Reich* complete the subjugation of the town-centre.

By 16 March, therefore, disaster loomed for the Soviets if Hausser's armour managed to capture the key communications hub of Belgorod, as this would threaten the lines of communication of the Soviet Central Front, then deployed west of Kursk. During 18–19 March, the SS-Panzer Corps, again with *Das Reich* in the vanguard, advanced north-east to capture Belgorod and east to reach the Donets, creating a German-held salient south of the Soviet bulge around Kursk. On 19 March, however, the temperature rose, thawing the frozen ground into a muddy morass that bogged down the panzers. In the ensuing weeks a lengthy pause descended over the front as both exhausted belligerents sought to rebuild their shattered forces for the summer 1943 campaign season. *Das Reich,* now deployed in reserve to the west of Kharkov, sorely needed this opportunity to absorb replacement equipment and manpower, for it had suffered 77 tank and 3500 personnel casualties in the previous six weeks of combat. During this period the division undertook some minor reorganization, providing cadres for the forming SS-Volunteer Brigade *Langemarck*. During this period the Germans began planning their summer 1943 offensive, codenamed 'Citadel', against the salient of Soviet-held territory that jutted west into the German frontline around Kursk.

Das Reich: 8./SS-Panzerregiment 2, November 1942– May 1943

Panzerkampfwagen VI Ausf E Tiger: 10
Panzerkampfwagen III Ausf N: 12

Until mid-1943, *Das Reich's* Tiger company fielded a mixture of Tiger I and Panzer III Ausf N. The latter were armed with the low-velocity 75mm (2.95in) L/24 close-support gun and were particularly useful in tackling hostile ant-tank guns, allowing the Tigers to concentrate on destroying enemy AFVs.

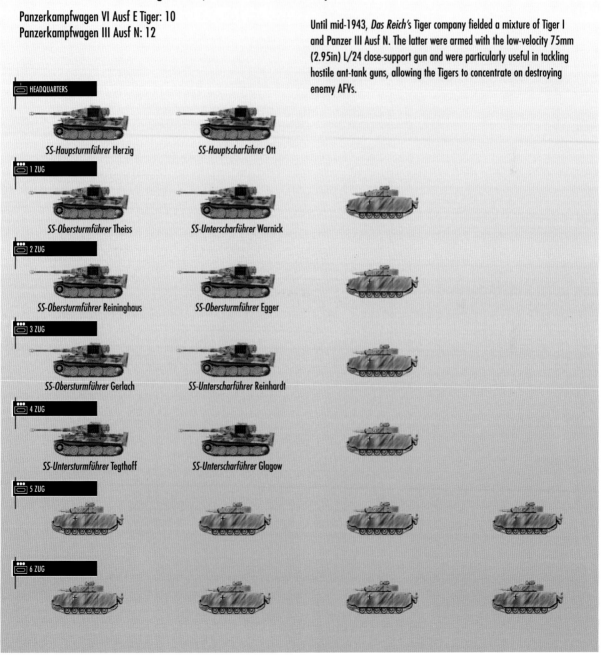

HEADQUARTERS

SS-Haupsturmführer Herzig   SS-Hauptscharführer Ott

1 ZUG

SS-Obersturmführer Theiss   SS-Unterscharführer Warnick

2 ZUG

SS-Obersturmführer Reininghaus   SS-Obersturmführer Egger

3 ZUG

SS-Obersturmführer Gerlach   SS-Unterscharführer Reinhardt

4 ZUG

SS-Untersturmführer Tegthoff   SS-Unterscharführer Glagow

5 ZUG

6 ZUG

Chapter 2
Building an Elite:
Das Reich,
1942–43

Between 1941 and 1945, almost four years of savage combat
on the Eastern Front saw the *Waffen-SS* expand from three divisions (*Leibstandarte,*
Das Reich and *Totenkopf*) with a total strength of roughly 60,000 men to 38 divisions
with almost a million troops. Such rapid growth inevitably led to a decline in quality in
many of the newer *Waffen-SS* formations, but despite suffering heavy casualties the
'old divisions' including *Das Reich,* managed to retain a genuine
elite status throughout the war.

In the case of *Das Reich,* this was partly due to the fact that it remained
an all-volunteer formation until mid-1944. But even during the last year of the war,
when it had to absorb large numbers of reinforcements compulsorily transferred from
the *Luftwaffe* and *Kriegsmarine,* its veterans were able to instil them with the
division's unique ethos and turn them into highly effective combat troops.

By the end of the war, the division's personnel had been awarded
69 Knight's Crosses and 151 German Crosses in Gold. *Das Reich* also claimed
the destruction of a total of 2000 Allied tanks, more than any
other German division.

OPPOSITE: Heinrich Himmler greets members of the *Das Reich* Panzer Regiment, seen here standing in front of Panzer VI
Tiger tanks, following their success in recapturing Kharkov in the spring of 1943.

II SS Panzer Corps, 1943

The formation that ultimately became II SS Panzer Corps was formed in June 1942 at *Truppenübungsplatz* Bergen-Belsen (Bergen-Belsen Troop Training Area) as *SS-Panzer-Generalkommando* under the command of *SS-Gruppenführer* Paul Hausser.

Initially, it comprised no more than a skeleton HQ staff, which was brought up to strength following its transfer to France to take command of *Leibstandarte*, *Das Reich* and *Totenkopf*, all of which had been withdrawn from the Eastern Front to refit. These refits entailed far more than the usual absorption of replacement personnel and equipment since all three formations were in the process of being upgraded from motorized divisions to exceptionally powerful panzergrenadier divisions. (At this stage, the formation was simply referred to as the SS Panzer Corps.)

The Soviet breakthrough at Stalingrad posed a serious threat to the very survival of Army Group South. As a result, in January 1943, Hitler ordered the Corps to be rushed to the Eastern Front to halt the Red Army's seemingly unstoppable advance, which had captured Kursk, Belgorod and Kharkov.

The Corps was von Manstein's most powerful formation and formed the spearhead of his brilliant

The crew of a Panzer III from the *Das Reich* division pose for a photograph as smoke from a recent engagement rises on the horizon.

'backhand' offensive, which smashed the Soviet advance in less than a month. The Germans inflicted 80,000 casualties on the Red Army besides recapturing Belgorod and Kharkov – it was a stunning reversal of fortune, marred only by von Manstein's failure to eliminate the Kursk salient before the spring thaw made further major operations impossible.

The *Waffen-SS* successes in what became known as the Third Battle of Kharkov impressed Hitler, and Himmler now seized his chance to press for an expansion of his 'private army'. At the end of March 1943, preliminary instructions were issued for the formation of two new formations based on cadres drawn from *Leibstandarte*:

■ SS Panzergrenadier Division *Hitlerjugend*

■ I SS Panzer Corps (commander-designate *SS-Obergruppenführer* Sepp Dietrich)

(The existing corps should have been designated I SS Panzer Corps, but became II SS Panzer Corps with effect from 1 July 1943. Although no formal reasons for this seem to have been

II SS Panzer Corps: AFV Strength, July 1943	
TYPE	STRENGTH
Panzer II	8
Panzer III	146
Panzer IV	168
Tiger I	42
Stug III	104
T-34	26
Wespe	36
Hummel	18

recorded, it is likely that Sepp Dietrich used his influence with Hitler to have his new command given the 'senior title'.)

These changes involved considerable upheaval within *Leibstandarte*. During the period from late March to early June, it lost the following personnel who were posted to form the cadres of the new formations:

■ 12 divisional staff officers

II SS Panzer Corps, 4 July 1943

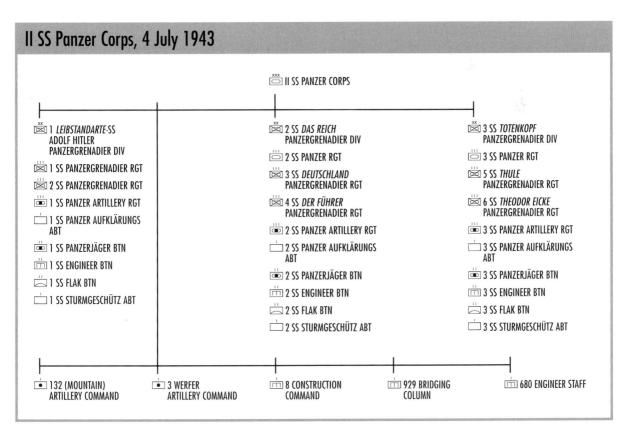

■ The commanders of both panzergrenadier regiments
■ The entire 1st SS Panzer Battalion
■ The battalion commander and one company from the reconnaissance battalion
■ A company commander from the assault gun battalion
■ An entire battalion from the artillery regiment
■ A battery commander from the flak battalion
■ Company commanders from signals and supply units.

Whilst it was relatively easy to fill the vacant staff officer posts, finding good replacement commanders for combat units was far more difficult. Fortunately, *Leibstandarte* had extremely able junior officers who were capable of filling these command posts, despite being markedly younger than their equivalents in army formations. However, it was not just officers who were required for these cadres – a total of 2500 other ranks were also transferred. By this stage of the war, there was no chance of finding so many suitable volunteer replacements and 2500 *Luftwaffe* personnel suddenly found themselves compulsorily transferred to *Leibstandarte*. Although most of them quickly settled in, the disruption had an adverse effect on the division and II SS Panzer Corps as a whole when they were in the midst of preparations for what Hitler described as 'the decisive battle of the war'.

In the circumstances, it was just as well that the repeated delays in launching Operation 'Citadel' gave Hausser time to bring his Corps to peak condition. Despite everything, by early July it was one of the most powerful and well-trained formations of its size on either side. Over 90 per cent of its 494 AFVs were operational when the offensive finally began on 5 July. In addition to its impressive armoured strength, it could field a total of 21 infantry battalions (including elements of 167th Infantry Division), 179 artillery pieces and 138 *Nebelwerfer* (salvo rocket launchers).

HIDDEN WEAKNESS

These impressive totals of AFVs tend to mask the two serious weaknesses in the German forces assembled for the Kursk offensive – the relative lack of infantry and artillery. The infantry had borne the brunt of German casualties on the Eastern Front since the beginning of Operation *Barbarossa* and although the number of infantry divisions had increased since June 1941, their official strengths had been steadily reduced. (Worse still, only a tiny minority of divisions could field anything like even these numbers.) This was to prove to be a crippling disadvantage in Operation 'Citadel', where the 23 infantry divisions committed were hopelessly overstretched, forcing panzer and panzergrenadier divisions to be misused for such tasks as flank protection when they should have been spearheading the offensive. (The situation was worsened by the fact that at least six of the 23 infantry divisions could not be released from other commitments until after the operation had started.)

LACK OF ARTILLERY

II SS Panzer Corps felt this lack of infantry from the beginning of the offensive because it was assigned just one such formation, the 167th Infantry Division. Even on 5 July, Hausser had to order *Totenkopf* to screen the Corps' right flank until the 167th could disengage from its left flank and move across to take over.

This was a frustrating scenario played out time and time again during the following week – just as it seemed that the Corps was on the verge of making a decisive breakthrough, another series of desperate Soviet attacks against its steadily lengthening flanks would force one or other of the panzergrenadier divisions to turn aside to deal with the threat.

The German weakness in artillery would also prove to be a crucial factor in the failure of the offensive – their total of 7417 guns and mortars was dwarfed by the Soviet total of 31,415. Considering that the attackers were faced with a fortified zone at least as formidable as anything seen in World War I, they were attempting a decisive breakthrough with less than 75 per cent of the artillery deployed in support of the German spring offensive of 1918.

Even for a well-equipped elite formation such as II SS Panzer Corps, this had the effect of reducing its ability to penetrate multiple Soviet defence lines without massive close air support. As the Red Air Force was making it increasingly difficult for the *Luftwaffe* to maintain even local air superiority, such support could rarely be guaranteed.

SS-Obergruppenführer Paul Hausser (7 October 1880 – 21 December 1972)

Hausser was born in Brandenburg an der Havel to a Prussian military family; his father Kurt Hausser was a major in the Imperial German Army. Paul entered the army in 1892, attending the cadet school in Köslin until 1896, when he transferred to the cadet academy Berlin-Lichterfelde, from which he successfully graduated in 1899. On 20 March 1899, he received his commission as a lieutenant and was posted to Infantry-Regiment 155 stationed at Ostrowo in Posen. In October 1903, he became the adjutant of the regiment's 2nd battalion, serving in this capacity for the next five years. Hausser was assessed to be an exceptionally promising young officer and was selected to attend the General Staff course at the Prussian Military Academy in Berlin in October 1908. He graduated in July 1911, thereafter holding a variety of General Staff posts throughout World War I. His ability was so outstanding that he was allowed to continue to serve in the tiny post-war German army (*Reichswehr*), retiring in January 1932 as a Lieutenant-General.

On his retirement, Hausser joined the right-wing veterans' organization *Stahlhelm*, becoming the head of its Brandenburg-Berlin chapter in 1933. Shortly afterwards, *Stahlhelm* was incorporated into the SA, and subsequently into the SS. In November 1934, Hausser was transferred to the *SS-Verfügungstruppe* and posted to *SS-Führerschule Braunschweig*. In 1935 he became Inspector of *SS-Junkerschule* and was promoted to *Brigadeführer* in 1936, with overall responsibility for training what was to become the *Waffen-SS*. He used the tactics pioneered by the highly mobile *Stosstruppen* (shock troops) of World War I as the basis for a rigorous training programme, which also emphasized the importance of sport, physical fitness and, above all, field craft. Hausser served in the Polish Campaign of 1939 as an observer with the mixed *Wehrmacht*/SS Panzer Division *Kempf*. In October 1939, SS-VT was formed as a motorized infantry division with Hausser in command. He led the division, later renamed 2nd SS Division *Das Reich*, throughout the French campaign of 1940 and in the early stages of Operation *Barbarossa*, until he was seriously wounded (losing his right eye) in October 1941. Following a period of convalescence, he commanded the newly formed SS-Panzer Corps (comprising *Leibstandarte*, *Das Reich* and *Totenkopf*), which was rushed to the Ukraine in January 1943 to reinforce Army Group South in the aftermath of the Stalingrad debacle. In mid-February, Hausser ignored Hitler's explicit order to hold Kharkov at all costs and withdrew his forces to avoid encirclement, only to recapture the city during von Manstein's counter-offensive the following month.

In July 1943, Hausser led the Corps (re-titled II SS Panzer Corps on 29 June 1943) throughout the Battle of Kursk. Hausser supervised the Corps' reorganization in France in January 1944, in which its former divisions were replaced by 9th and 10th SS Panzer Divisions (*Hohenstaufen* and *Frundsberg*). He continued to command the Corps with distinction in Russia in early 1944 and subsequently in the early stages of the Normandy campaign until he was again wounded. Paul Hausser was promoted to *Oberstgruppenführer und Generaloberst der Waffen-SS* in August 1944 and subsequently commanded Army Group G from 28 January to 3 April 1945. He ended the war on *Generalfeldmarschall* Albert Kesselring's staff.

2nd SS Panzergrenadier Division *Das Reich*, 1943

In June 1942, the exhausted and battered SS Division (motorized) *Das Reich* was withdrawn from the Eastern Front and sent to France to refit.

In normal circumstances, this would have involved little more than straightforward re-equipment and the training of reinforcements, but on this occasion, the process was far more complex as *Das Reich* was in the throes of conversion to a panzergrenadier division.

REORGANIZATION

On 9 November 1942, the reorganization was officially completed and the division formally became 2nd SS Panzergrenadier Division *Das Reich*. Within a matter of weeks, its routine training and occupation duties were interrupted as it received orders to standby to participate in Operation *Lila,* the seizure of the Vichy French fleet at Toulon. The operation had been planned in response to the Allied invasion of French North Africa (Operation 'Torch') on 8 November 1942, which aroused Hitler's fear that the fleet might defect to General de Gaulle's Free French Forces.

Das Reich formed two panzer *kampfgruppen* (armoured battlegroups) for the operation, which were to be supported by its motorcycle battalion. These units went into action on 27 November in conjunction with elements of 7th Panzer Division and *Kriegsmarine* detachments, which were to take control of the French vessels. Whilst the naval dockyard was occupied with scarcely a shot being fired, virtually all the ships were scuttled by their crews. (According to some accounts, one of the *kampfgruppen* came under fire from the AA machine-guns of the battleship *Strasbourg* and hastily retreated when her secondary 130mm/5.1in guns were trained on them, giving her crew sufficient time to set scuttling charges.)

Following this slightly farcical operation, the division returned to routine duties until the disaster at Stalingrad forced its hasty redeployment to the Eastern Front in January 1943. *Das Reich*, together with *Leibstandarte* and *Totenkopf*, played a key role in defeating the Soviet winter offensive, Operation 'Star', which had been undertaken to recapture Belgorod and Kharkov. However, this success came at a high price – the division sustained almost 4500 casualties, about 25 per cent of its total strength. The lull in the fighting from the end of March allowed *Das Reich* to spend the next two months rebuilding its battered units – especially the panzergrenadier and *pionier* (assault engineer) battalions, which almost invariably took the heaviest losses. The replacement personnel who arrived throughout the period were a very mixed group, ranging from wounded veterans returning after convalescent leave to ill-prepared raw recruits who needed considerable additional training to give them a reasonable chance of survival in combat.

Besides these reinforcements, there was a further group that had a shadowy, semi-official existence – the so-called Russian *Hilfswilligen* (or *Hiwis*, volunteer helpers), the majority of whom were employed in rear areas in posts such as sentries, drivers, store keepers and kitchen staff. However, significant numbers were also deployed in combat roles. The process had begun as early as July 1941 when the 134th Infantry Division began openly to enlist Russians. In September 1941 Hitler officially sanctioned the recruitment, but specified that all such volunteers could serve only as unarmed auxiliaries – a ruling that was almost universally ignored. By the spring of 1942, there were an estimated 200,000 *Hiwis* serving in the rear areas of the German armies, and by the end of the same year their number had risen to something in the order of 250,000.

Das Reich Division: AFV Strength, July 1943	
TYPE	STRENGTH
Panzer III	53
Panzer IV	27
Tiger I	11
Command Tanks	8
Stug III	21
T-34	16

SS-Obergrüppenfuhrer Walter Krüger (27th February 1890 – 22nd May 1945)

Walter Krüger was born in Strasbourg, Alsace-Lorraine (then part of the German Empire). Like his father, he decided upon an army career and was commissioned as a second lieutenant in the 110th Fusilier Regiment. During his service in World War I, he was wounded twice and awarded both classes of the Iron Cross. Immediately after the war, he joined the *Freikorps* and fought in East Prussia before working as a bank clerk and estate manager.

He joined the Austrian SA in 1933, transferring to the *SS-Verfugungstruppe* in April 1935 with the rank of *SS-Obersturmbannführer.* Kruger formed the *SS-Standarte Germania and* subsequently served as an instructor at the *SS-Junkerschule* at Bad Tölz.

During much of 1940, he served with the *Polizei* Division, winning clasps to his Iron Cross, before returning to administrative duties. In August 1941, he was appointed as commander of the *SS-Polizei* Division and was awarded the Knight's Cross on 14 December 1941 in recognition of the division's performance in the siege of Leningrad. After a further brief period of administrative duties, Kruger replaced the wounded Herbert Vahl as the commander of *SS-Panzergrenadier-Division Das Reich* in March 1943. He commanded the division with great skill throughout Operation 'Citadel' and was awarded the Oak Leaves to his Knight's Cross in

September 1943. After relinquishing the command of the division in March 1944, he was promoted to *SS-Obergrüppenfuhrer* and took command of the newly formed VI *Waffen-Armee Korps der SS (lettisches)* (VI Latvian SS Corps), which retreated into the Courland Pocket with the rest of Army Group North in October 1944. Kruger was awarded the Swords to his Knight's Cross in January 1945 and his Corps played a key role in defeating repeated Soviet offensives until ordered to surrender on 8 May 1945. He joined a group of officers attempting to escape to East Prussia, but was trapped by Soviet forces and committed suicide to avoid capture.

All three divisions in II SS Panzer Corps had substantial numbers of *Hiwis* at the beginning of Operation 'Citadel' (*Das Reich*, 1576; *Leibstandarte*, 1238; and *Totenkopf*, 1147), even though they did not approach the maximum authorized level of 15 per cent of divisional strength. By this stage of the war, the *Hiwis'* conditions of service had been formalized, and the vast majority were properly uniformed with their rations and pay at near parity with those of German troops. Even for the *Waffen-SS,* the practicalities of campaign life rapidly made Himmler's rhetoric about Slavic *Untermenschen* completely irrelevant. In a typical and completely

unofficial move, *Der Führer's* commander, *Sturmbannfuhrer* Stadler, had established a regimental intelligence cell, which included six *Hiwis* who listened in to Red Army radio traffic and provided important tactical intelligence throughout the campaign.

POWERFUL UNITS

By the beginning of July 1943, *Das Reich*, together with *Leibstandarte* and *Totenkopf*, had become three of the most powerful divisions in the entire German ORBAT, with markedly greater resources than standard army panzer divisions.

However, even such powerful formations were to prove unable to fulfil the enormous demands placed on them by Hitler in Operational Order No. 6 of 15 April 1943, the definitive directive for the Kursk offensive:

'I have decided to launch Operation 'Citadel' as the first of this year's offensives, as soon as the weather permits. The offensive is of decisive importance. It must succeed quickly and completely. It must give us the initiative for the spring and summer of this year. Therefore all preparations are to be carried out with the greatest care and energy. The best formations, the best weapons, the best commanders and large quantities of ammunition are to be deployed at the decisive points. Every officer and every man must be totally convinced of the decisive importance of this offensive. The victory of Kursk must be a beacon to the world.

I therefore order:

(1) The objective of the offensive is the encirclement of the enemy forces deployed in the Kursk area by means of concentrated, coordinated, merciless and rapid thrusts by one attacking army each from the areas of Belgorod and south of Orel, and to annihilate them by concentric attack. In the course of the offensive a shortened front, which will liberate forces for use elsewhere is to be established along the line Nezhega – Korocha (exclusive) – Skorodnoye – Tim – east of Shchigry – Sosna (exclusive). ...

(2) It is vital:

a. to maintain the element of surprise as far as possible, and in particular to keep the enemy guessing about the timing of the offensive.

b. to concentrate the attacking forces on the narrowest possible front, so that they can achieve an

Units in Panzer and Waffen SS Panzergrenadier Divisions Compared

TYPE	PANZER DIVISION	WAFFEN-SS
Panzer Battalions	2	2
Tiger Company	–	1
Motorized Infantry Btn	3	5
Armoured Infantry Btn	1	1
Light Artillery Btn	1	2
Heavy Artillery Btn	1	1
SP Artillery Btn	1	1
Reconnaissance Btn	1	1

ORBAT: 2 SS Panzergrenadier Division *Das Reich*

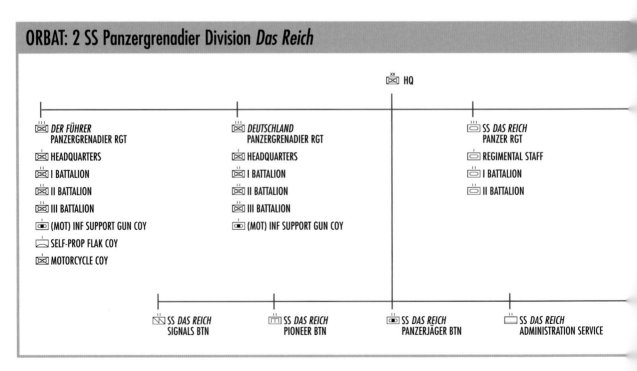

HQ

DER FÜHRER PANZERGRENADIER RGT
- HEADQUARTERS
- I BATTALION
- II BATTALION
- III BATTALION
- (MOT) INF SUPPORT GUN COY
- SELF-PROP FLAK COY
- MOTORCYCLE COY

DEUTSCHLAND PANZERGRENADIER RGT
- HEADQUARTERS
- I BATTALION
- II BATTALION
- III BATTALION
- (MOT) INF SUPPORT GUN COY

SS DAS REICH PANZER RGT
- REGIMENTAL STAFF
- I BATTALION
- II BATTALION

SS DAS REICH SIGNALS BTN

SS DAS REICH PIONEER BTN

SS DAS REICH PANZERJÄGER BTN

SS DAS REICH ADMINISTRATION SERVICE

overwhelming local superiority (of all arms) allowing them to break through the enemy in a single move and seal off the pocket.

c. to ensure that the attacking forces are reinforced by units brought up from the rear to cover their flanks so that attacking forces can concentrate all their efforts on the advance.

d. to attack the pocket from all sides, not giving the enemy any respite, and thus hasten his annihilation.

e. to attack so swiftly that enemy is prevented from breaking contact or bringing up strong reserves from other fronts.

f. to rapidly establish a new front to free forces, particularly mobile formations, for further operations as soon as possible …

(8) Because of the reduced scale of the offensive compared to earlier operations … the attacking forces are to leave behind all vehicles and impedimenta which are not absolutely necessary for the attack. … every commander must see that he takes with him only that which is absolutely essential for battle. Corps commanders and divisional commanders must ensure the execution of this directive. Strict traffic control is to be established. …

(12) The ultimate objectives of the operation are:
a. The movement of the boundary between Army Groups South and Centre to line Konotop (South) – Kursk (South) – Dolgoye (Centre).
b. The transfer of Second Army HQ, with three corps HQs and nine infantry divisions, as well as other army troops yet to be designated, from Army Group Centre to Army Group South.
c. The deployment of three further infantry divisions from Army Group Centre, to be available for deployment by OKH north-west of Kursk.
d. The withdrawal of the majority of armoured formations from this front for deployment in other theatres. …'

Das Reich fought as hard as any formation could to try to fulfil Hitler's directive, but finally had no choice but to admit failure – as the divisional history noted, it was:

'… clear that *Zitadelle* could not succeed. For on both sides of the northern and southern flanks the German advances had not gained the ground expected of them and there were still 130 kilometres between the pincers of Kempf's and Hoth's armies; 130 kilometres of trenches, minefields and Russian armour.'

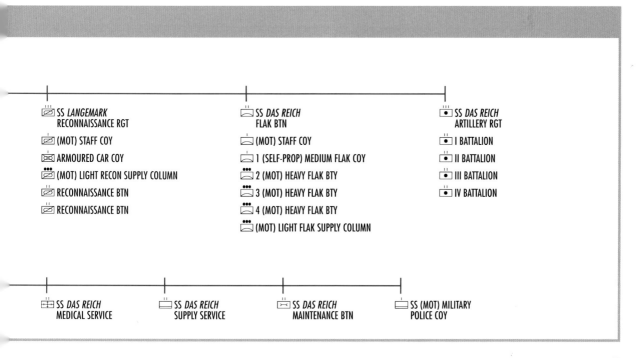

Der Führer SS Panzergrenadier Regiment, 1943

Following the German annexation of Austria, *SS-Standarte Der Führer* was formed March 1938 in Vienna as *SS-Standarte 3*. The unit was raised from Austrian volunteers with a cadre drawn from *SS-Standarte Deutschland*. It was renamed *SS-Standarte Der Führer* later in 1938, when it also received its regimental standard.

It took part in the occupation of Czechoslovakia before serving as a guard regiment in Prague – the *Wach-Regiment des Reichsprotektors von Böhmen und Mähren*. The unit was not committed to action in the Polish campaign as it was not fully trained, but was sent to garrison a sector of the Westwall (Siegfried Line). In October 1939 it was combined with the *Deutschland* and *Germania* regiments to form the *SS-Verfügungs* Division (later renamed *Das Reich*). *Der Führer* retained much of its Austrian character throughout the war, although this was steadily diluted as replacement personnel were increasingly drawn from across the Reich.

It is probably fair to say that this meant that the regiment was the focus of loyalty for most of its personnel – they would tend to think of themselves primarily as members of *Der Führer*, with the division being a more remote entity. This was reinforced by the intensive pre-war and early wartime training of the *Waffen-SS*. The training programme was largely devised by the ex-*Reichswehr* General Paul Hausser, who was appointed to the position of Inspector of *SS-Verfügungstruppen* in October 1936, and Felix Steiner, a former stormtrooper from Worl War I. Steiner was primarily responsible for designing many of the revolutionary and highly realistic training programmes that emphasized practical and classroom training, and put less focus on the traditional army 'square-bashing'. As soon as the men were familiar with their weapons, the main focus of the course shifted to infantry assault tactics. Instructors put great emphasis on aggression, constantly stressing speed and ferocity in attack in order to win quickly and minimize casualties. All ranks took part in the same daily sporting activities, which produced a high level of unit fitness and also '...fostered among officers, NCOs, and enlisted men a sense of fellowship and mutual respect generally unknown in the Army'.

ORBAT: *Der Führer* Panzergrenadier Regiment

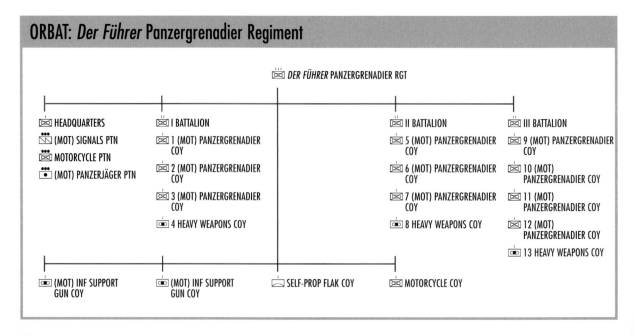

Der Führer SS Panzergrenadier Regiment: Equipment

UNIT	EQUIPMENT	STRENGTH	UNIT	EQUIPMENT	STRENGTH
Regimental HQ	Kübelwagen	3	Armoured Infantry Battalion		
	Kettenkrad	4	HQ	Motorcycle	3
HQ Company	Kettenkrad	1		Kettenkrad	2
	Motorcycle	1		Kübelwagen	3
	SdKfz 251/3	1		SdKfz 251/3	4
Signals Ptn	Motorcycle	1		SdKfz 251/8	4
	Kettenkrad	1	Company (x3)		
	SdKfz 251/3	6	HQ	Motorcycle	2
	SdKfz 251/11	3		Kübelwagen	1
Reconnaissance Ptn	Motorcycle	3		Kettenkrad	3
	Kübelwagen	1		SdKfz 251/3	2
	Kettenkrad	15	Infantry Ptn (x3)	SdKfz 251/10	1
Panzerjäger Ptn	Kettenkrad	1		SdKfz 251/1	3
	SdKfx 251/10	1	Heavy Ptn	SdKfz 251/10	1
	SdKfx 251/1	1	HMG Section	SdKfz 251/1	3
	Maultier/PaK 40	3	Mortar Section	SdKfz 251/2/80mm mortar	2
	LMG	2	Heavy Company		
Panzergrenadier Battalion (x2)			HQ	Motorcycle	2
Battalion HQ	Motorcycle	7		Kettenkrad	2
	Open Blitz Type A	4		SdKfz 251/3	2
	Kübelwagen	5		SdKfz 11	2
Company (x3)			Light Infantry Gun Ptn	SdKfz 251/4	2
HQ	Motorcycle	8		75mm leIG (towed)	2
	Kübelwagen	2	Infantry Gun Ptn	SdKfz 251/3	1
Infantry Ptn (x3)	Kublewagen	2		SdKfz 251/1	1
	Opel Blitz Type A	3		SdKfz 251/9	6
Heavy Ptn	Kublewagen	2	Panzerjäger Ptn 1	SdKfz 251/1	1
	Opel Blitz/MG42 MG	2		SdKfz 251/4	4
Mortar Section	Opel Blitz/80mm mortar	2		75mm PaK 40 (towed)	3
	LMG	18	Panzerjäger Ptn 2	SdKfz 251/1	1
	HMG	4		SdKfz 251/4/2.8cm sPzB41	3
	80mm mortar	2	SP Heavy Infantry Gun Coy	Kettenkrad	5
	Flamethrowers	2		Kübelwagen	4
Heavy Weapons Company				Maultiers	9
HQ	Motorcycle	5		SdKfz 251/3	1
	Kübelwagen	4		SdKfz 251/1	3
	Opel Blitz Type A	2		SdKfz 251/11	1
	Maultier	3		150mm sIG PzKpwg 38(t)	6
Mortar Ptn	Opel Blitz/80mm mortar	6	Self-propelled Flak Company	Motorcycle	12
Panzerjäger Ptn	Kübelwagen	1		Kübelwagen	7
	Maultier	1		Maultier	3
	Maultier/PaK 40	3		Kfz 17 radio car	5
Panzerjäger Ptn	Kübelwagen	1		SdKfz 10	1
	Maultier	1		SdKfz 10/5 20mm cannon	12
	Kfz 81/2.8cm sPzB41	3	Motorcycle Company	HMG	4
Infantry Support Gun Sec	75mm leIG	4		LMG	18
Flamethrower Section	SdKfz 251/16	2		80mm mortar	2
				75mm PaK 40	3

Although it seems that Steiner had little time for the extremes of Nazi ideology, he had to pay lip-service to Himmler's insistence on incorporating it into the *Waffen-SS* training programme. As a result, lectures on Nazi policies, SS philosophies and especially the theories of Aryan racial superiority were held at least three times a week. Steiner, though, made sure that his men were focused on increasing their combat effectiveness rather than becoming experts in the field of pseudo-scientific racial science. As a later commentator put it:

'He believed firmly in the creation of élite, highly mobile groups whose training put the emphasis on individual responsibility and military teamwork rather than mindless obedience.'

The two *Junkerschule* at Bad Tölz and Braunschweig, 'a cross between the Spartan hoplites and the Guards Depot at Caterham', consistently produced high-quality junior officers, despite the fact that 40 per cent of cadets had no more than a basic education. Early in the war, there were indications that the training had its limitations: the high level of aggression instilled in all ranks of the *Waffen-SS* led to reckless attacks and horrendous casualty lists. However, these lessons were soon absorbed and the training was now tempered by an acknowledgement of the need to avoid unnecessary casualties.

WARTIME CHANGES

By mid-1943, *Der Führer,* in common with the rest of the frontline *Waffen-SS,* was increasingly regarding itself as 'the fourth arm of the *Wehrmacht*', with casual day-to-day usage of army ranks rather than official SS titles and a growing incidence of displays of contempt for the Nazi party hierarchy, including the *Reichsführer-SS* himself. (As early as November 1942, Himmler had reprimanded Hausser for his tolerance of outspoken criticism of all aspects of the SS organization.)

Such tensions exploded in the aftermath of the *Waffen-SS* recapture of Kharkov in March 1943. The city had been an important administrative centre for the German-occupied Ukraine, but the high-ranking Nazi administrators had hastily fled before it fell to the Red Army on 16 February. As soon as the area was safe, these officials returned. One *Waffen-SS* panzergrenadier recalled their arrival in a fleet of Mercedes, accompanied by their blonde girlfriends, and their attempt to reclaim their former quarters from him and his comrades. This resulted in several beaten-up bureaucrats.'

Significantly, the one person exempt from the general contempt expressed for the Nazi party was Hitler himself. This was largely due to the general perception that he had earned respect for his service as a brave frontline soldier during World War I.

Regimental HQ Company

UNIT	EQUIPMENT	STRENGTH
HQ Company	Kettenkrad	1
	Motorcycle	1
	SdKfz 251/3	1
Signals Ptn	Motorcycle	1
	Kettenkrad	1
	SdKfz 251/3	6
	SdKfz 251/11	3
Reconnaissance Ptn	Motorcycle	3
	Kübelwagen	1
	Kettenkrad	15
Panzerjäger Ptn	Kettenkrad	1
	SdKfx 250/10	1
	SdKfx 251/1	1
	Maultier/PaK 40	3
	LMG	2

Der Führer Panzergrenadier Regiment: HQ elements

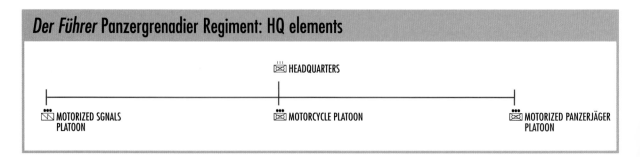

HEADQUARTERS

MOTORIZED SGNALS PLATOON

MOTORCYCLE PLATOON

MOTORIZED PANZERJÄGER PLATOON

As combat troops, they could identify with him far more readily than with the bureaucratic figure of Himmler.

Hitler's faith in them was matched by their determination to succeed in missions that the army believed were impossible. *Grüppenfuhrer* Kruger's message to the officers and men of *Das Reich* on the eve of Operation 'Citadel' appealed both to their inherent *Waffen-SS* comradeship and their personal sense of loyalty to Hitler:'That indestructible bond of comradeship at the front … and the spirit of the *Schutzstaffel* binds me to every one of you. As always, the *Führer* counts on our dash and determination!'

Regimental HQ Company

Kettenkrad: 18	SdKfz 251/3: 7
Kübelwagen: 1	SdKfz 251/11: 3
Motorcycle: 5	SdKfz 250/10: 1
Maultier/PaK 40: 3	SdKfz 251/1: 1

The regimental HQ company was involved primarily in communications and administrative tasks, although it did include a gun section containing three PaK 40 anti-tank guns. These would be assigned to other units where needed.

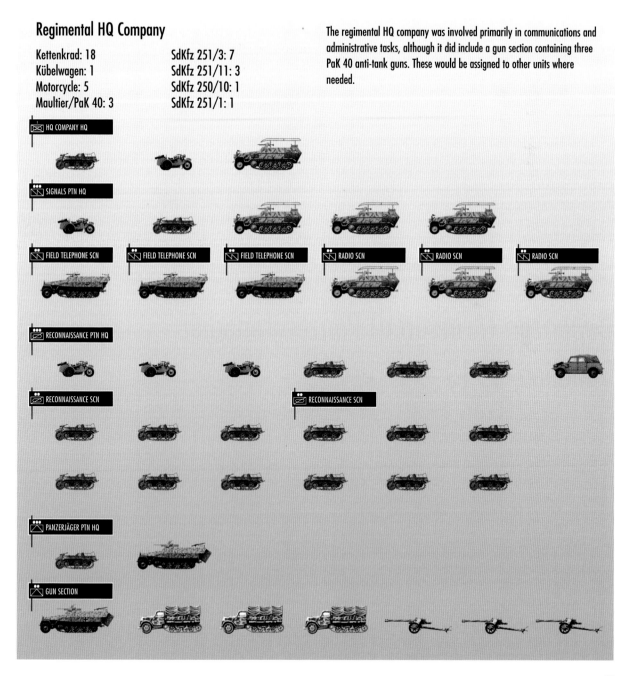

Volkswagen Kübelwagen

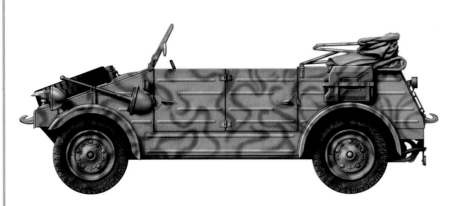

■ CREW

■ SPECIFICATIONS

Weight: 715kg (1580lb)
Length: 3.74m (12ft 3in)
Width: 1.6m (5ft 3in)
Height: 1.65m (5ft 5in)
Engine: 17.15kW (23hp)/18.64kW
 (25hp) air-cooled flat four petrol
Speed: 80km/h (50mph)
Range: 600km (375 miles)

The Kübelwagen (a title based on Kübelsitzwagen – 'bucket seat car') was designed by Ferdinand Porsche as a military derivative of the Volkswagen Beetle. The prototype, designated Type 62, underwent military trials in November 1938; despite its 4x2 configuration, the vehicle proved to have a surprisingly good cross-country performance due to its light weight and ZF self-locking differential. Further development of the Type 62 took place during 1939, including a more angular body design; and pre-production models were field-tested in the Polish campaign. The additional modifications incorporated as a result of these combat trials led to the vehicle's re-designation as the Type 82. Full-scale production of the Type 82 Kübelwagen began in February 1940 and continued throughout the war; total production amounted to 50,435 vehicles. The only significant change to the design was made in March 1943 when a 1131cc engine developed for the Schwimmwagen replaced the original 985cc unit.

Gepanzert (Armoured) Panzergrenadier Battalion, 1943

The armoured panzergrenadier battalion was one of the most powerful wartime units of its size, combining high mobility with exceptional firepower.

By 1943, these battalions had developed a sophisticated tactical doctrine for a wide range of operations. Although they never had the benefit of fully-tracked armoured personnel carriers (APCs), the capabilities of their halftracks were exploited in a wide range of skilful operations.

EQUIPMENT:
THE SDKFZ 251
The SdKfz 251 halftrack was the key item in the panzergrenadier unit's inventory. By 1943, it had spawned a plethora of variants and more were being added until the last months of the war.

Development of the SdKfz 251 began in 1937 and the first production vehicles were issued to the 1st Panzer Division in 1939. Only small batches of the initial Ausf A and Ausf B versions were completed in 1939/40 because production of the complex, multi-faceted armoured hulls was a particularly slow business.

Matters improved with the introduction of the Ausf C in late 1940. This introduced a single sloping front plate for the engine compartment, which improved protection and simplified construction. The side-mounted engine cooling flaps were replaced by more efficient armoured vent covers for better cooling. Combat experience had shown that the front tyres were prone to damage from fouling the mudguards when crossing rough terrain. The mudguards of the Ausf C were raised to solve this problem and minor changes were made to the seating and stowage arrangements. The radio was also moved to a more accessible position under the front armour in front of the commander's seat.

These modifications produced a very good vehicle with a long track run using the characteristic 'slack track' design without return rollers. This lowered ground pressure and improved traction, giving the SdKfz 251 better cross-country performance than most other nations' halftracks. The overlapping and interleaved main road wheels common to virtually all German halftracks of the period gave a smooth ride, but were prone to jamming with ice or frozen mud in the extreme conditions of Russian winters.

Exhausted panzergrenadiers from the Das Reich Division ride forward in an SdKfz 251/1, supported by StuG assault guns. The halftrack is stacked with extra ammunition for the forward-mounted MG42.

SdKfz 251 Types

The types most commonly found in the panzergrenadier battalions were:

■ SdKfz 251/1: *Schützenpanzerwagen*. Standard armoured personnel carrier (APC).

■ SdKfz 251/2: *Schützenpanzerwagen (Granatwerfer)*. 81mm Mortar carrier

■ SdKfz 251/3: mittlere *Kommandopanzerwagen (Funkpanzerwagen)*. Communications vehicle, fitted with extra radio equipment for the command role.

■ SdKfz 251/4: *Schützenpanzerwagen für Munition und Zubehör des leIG18*. Gun-towing tractor, initially for use with the 7.5cm *leichtes Infanteriegeschütz 18*. Later used with the 50mm PaK 38, 75mm PaK 40 and 10.5cm leFH 18 Light Field Howitzer.

■ SdKfz 251/8: *Krankenpanzerwagen*. Armoured ambulance.

■ SdKfz 251/9: *Schützenpanzerwagen* (7.5cm KwK37). Equipped with a 75mm L/24 low velocity gun, nicknamed 'Stummel' (stump)

■ SdKfz 251/10: *Schützenpanzerwagen* (3.7cm PaK). Equipped with a 37mm PaK 36 anti-tank gun mount. Platoon commander's variant.

■ SdKfz 251/11: *Fernsprechpanzerwagen*. Telephone line layer.

Whilst the design offered good protection from small arms fire and shell splinters, all SdKfz 251/1s had open-topped troop compartments, which were vulnerable to a wide range of weapons such as air-burst artillery fire, mortars and grenades. This was a weakness shared with all other Axis and Allied wartime APCs – the first fully-enclosed designs entered service only in the 1950s.

The SdKfz 251/1 APC was armed with two MG34s or MG42s, one on a shielded mount above the driver's compartment, with the second on an AA mount above the rear doors. (This latter weapon was usually removed to accompany the panzergrenadiers when they dismounted. Whenever possible, their APC would give additional fire support from its front machine-gun.) In the pursuit, or when attacking weak enemy forces, the panzergrenadiers would not dismount; instead, they lay down suppressive fire from the vehicle – and in such situations, its open-topped design offered a positive advantage.

For its day, the basic SdKfz 251/1 was a highly sophisticated APC, although it fell short of the ideals propounded by Heinz Guderian, who had been largely responsible for the development of the *Panzerwaffe*. He had been convinced of the necessity of producing an all-arms armoured force and, as he later wrote in his book *Panzer Leader*:

'In this year (1929) I became convinced that tanks working on their own or in conjunction with infantry could never achieve decisive importance. My historical studies; the exercises carried out in England and our own experience with mock-ups had persuaded me that the tanks would never be able to produce their full effect until weapons on whose support they must inevitably rely were brought up to their standard of speed and of cross-country performance. In such formation of all arms, the tanks must play the primary role....'

Guderian wanted fully-tracked APCs and self-propelled artillery that could not only match the tank's mobility but which would have enough armoured protection to operate alongside them. Reluctantly, he had to accept that the halftrack was the best solution that could be produced in any sort of useful numbers by the German war industries. (Prototypes of the *Kätzchen*, a fully-tracked APC based on the hull of the Panzer 38(t), were built in 1944/45, but the type never entered service.)

Panzergrenadier Battalion: Personnel

UNIT	OFFICERS	MEN
Battalion Headquarters	6	20
Communications Platoon	1	22
Battle Train I	–	25
Battle Train II	3	44
Rations Train	1	6
Baggage Train	–	4
Heavy Company	4	129
Company HQ	1	15
Maintenance Detachment	–	13
Battle Train	–	11
Baggage Train	–	4
Anti-tank Platoon	1	31
Infantry Gun Platoon	1	24
Cannon Platoon	1	34
3 x Rifle Companies	3	217
Company HQ	1	17
Tank Destroyer Section	–	10
Maintenance Detachment	–	9
Battle Train	–	12
Baggage Train	–	4
Heavy Platoon comprised of:		
Platoon HQ	1	10
Mortar Section	–	15
Cannon Section	–	8
2 Heavy Machine Gun Sections, each	–	11
3 Rifle Platoons, each comprised of:		
Platoon HQ	1	6
3 Rifle Squads, each comprised of	–	10
Total Strength	24	896

PANZERGRENADIERE TACTICAL DOCTRINE

The training manual *Führung und Kampf der Panzergrenadiere* (Panzergrenadier Command & Combat Operations) described their functions:

'Mechanized *panzergrenadiere* are the *sturmtruppen* (armoured assault troops) of the panzer and panzegrenadier divisions ... Together with the panzers they form a close combat team ... The primary tasks of the panzergrenadiere are to:

■ Support panzers in the attack by eliminating anti-tank guns.

■ Protect the panzers from enemy infantry anti-tank teams.

■ Clear, occupy and hold territory gained by the panzers.

■ Rapidly exploit successful panzer operations.
■ Provide security for panzer units in their assembly points, during lulls in combat and whilst on the move.'

The Spanish Civil War and the campaigns of 1939/40 showed that without their own supporting infantry, tanks were always vulnerable to attack by determined enemy infantry. This vulnerability was

SdKfz 251/3 Ausf B

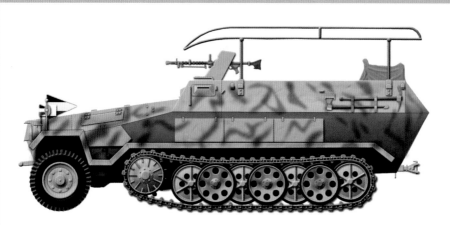

■ CREW

■ SPECIFICATIONS

Weight: 8.5 tonnes (8.36 tons)
Length: 5.8m (19ft)
Width: 2.1m (6ft 11in)
Height: 1.75m (5ft 9in)
Engine: 74.5kW (100hp) Maybach HL 42 TUKRM 6-cylinder petrol
Speed: 53km/h (33mph)
Range: 300km (186.4 miles)
Armament: 2 x 7.92mm (0.31in) MG34 or MG42 machine-guns

This designation covered a plethora of radio and command vehicles, most of which were fitted with the distinctive frame antenna for the FuG8 radio until well into 1943. Depending on the precise role, a wide variety of other radios were also fitted, whilst vehicles used by senior commanders often carried Enigma coding equipment.

SdKfz 251/7 Ausf C

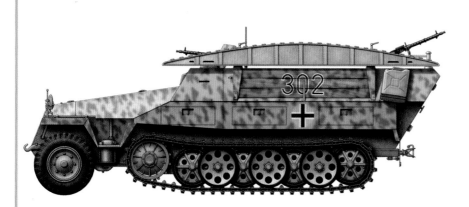

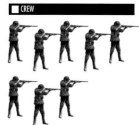

■ CREW

■ SPECIFICATIONS

Weight: 8.9 tonnes (9.1 tons)
Length: 5.80m (19ft)
Width: 2.10m (6ft 11in)
Height: 2.70m (8ft 10in)
Engine: 74.5kW (100hp) Maybach HL42 TUKRM
Speed: 50km/h (31mph)
Range: 300km (186 miles)
Armament: 1/2 x 7.62mm (0.31in) MGs

The engineer battalion attached to motorized and Panzergrenadier divisions had one halftrack company and two motorized companies. The SdKfz 251/7 carried light bridges, mines and other specialist combat engineer material.

increased if the infantry had good anti-tank weapons, but even poorly-equipped troops could pose a real threat if they had the necessary courage. Finnish infantry anti-tank teams destroyed about 1600 Soviet AFVs during the Winter War of 1939/40, mostly using Molotov cocktails by exploiting the vulnerability of ill-supported tanks in thickly forested areas.

When the panzers were fighting their way through defence lines or moving across terrain that provided the enemy with good cover, they needed their accompanying infantry to go in first to clear the way. Thus the panzergrenadiers might very often have to fight like conventional infantrymen. Conversely, in a fast-moving advance they might be carried by halftracks, trucks, motorcycles or, in extreme circumstances, on the panzers themselves, ready to dismount and engage anything that slowed

the tempo of the advance. Whenever the panzers bypassed areas of stiff enemy resistance, it was the job of the panzergrenadiers to eliminate these pockets, or at least to seal them off until the conventional infantry divisions could catch up and take over the task.

HALFTRACK TROOPS

Although the classic image of the panzergrenadier is closely associated with the SdKfz 251 halftrack armoured personnel carrier, there were never enough of these vehicles to fully equip all panzergrenadier formations. The concept of a carrier-borne attack into the heart of the enemy's defences accompanying the tanks was the ideal, but the reality was very different as most panzergrenadiers were transported in vulnerable trucks and on motorcycles, which required careful

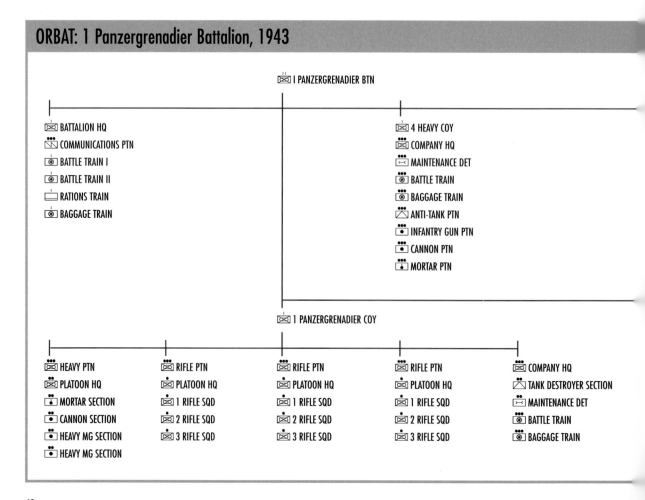

ORBAT: 1 Panzergrenadier Battalion, 1943

handling in order to survive on the battlefield. Therefore instead of driving into the midst of the enemy position, the motorized panzergrenadiers normally debussed at a forming-up point or start line away from the enemy's line of sight. They then attacked in the conventional manner of infantry supporting tanks. The key tactical advantage was that because of their motorization, they could be brought into battle as soon as they were needed.

It was only at the time of Operation *Barbarossa* in 1941 that SdKfz 251s became available in sufficient numbers to equip full panzergrenadier battalions, allowing the Germans to experiment with fighting directly from their halftracks. Although the SdKfz 251s provided reasonable protection against small arms fire, they had a maximum of 13mm (0.5in) of armour. This left them vulnerable to even the smallest calibre anti-

tank weapons, and they suffered accordingly. After taking heavy losses when attempting to deploy them alongside the panzers in combat, the Germans learned to debus at least 400m (1312ft) before enemy positions. Nonetheless, under certain tactical conditions, the halftrack was found to provide a useful firing position.

At the lowest level, the basic panzergrenadier unit was the *gruppe*, or squad, usually about 12 men mounted in a half-track or often a truck. The squad was led by a squad leader, usually a junior NCO, who was armed with an MP40 submachine-gun and was responsible for the squad to the platoon commander. On the move, he also commanded the vehicle and fired the front-mounted MG34 or MG42.

His rifle-armed assistant was normally a lance-corporal and would take command of the half

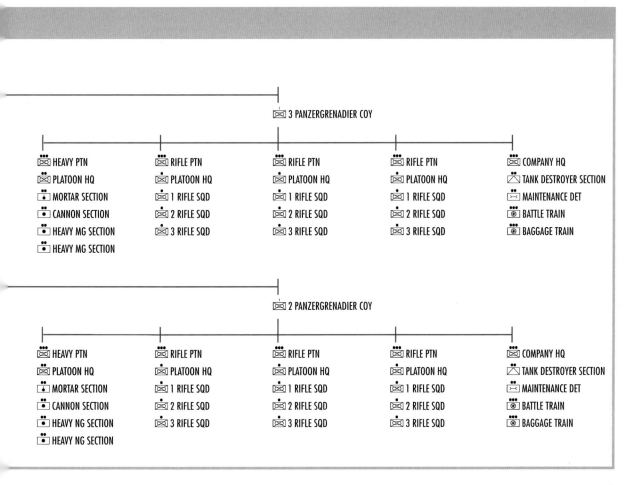

Armoured Infantry Battalion, Battalion HQ & Heavy Company: Equipment

Motorcycle: 5
Kübelwagen: 3
Kettenkrad: 4

SdKfz 251/3: 7
SdKfz 251/8: 2
SdKfz 251/11: 2

SdKfz 251/1: 3
SdKfz 251/4 tractor: 11
SdKfz 251/9: 6

PaK 40 AT gun: 3
75mm leIG howitzer: 2
2.8cm sPzB41: 3

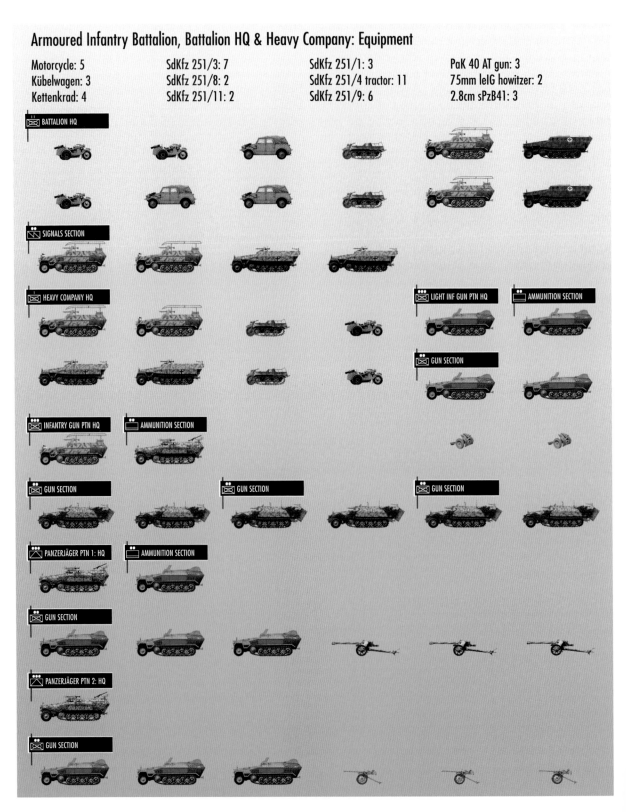

squad if the unit was divided. The squad contained two light machine-gun teams, each of two men, four rifle-armed infantrymen, plus the driver and co-driver. The driver was also responsible for the day-to-day vehicle maintenance and was expected to remain with his vehicle. The squad entered the SdKfz 251 through the rear doors, with the squad leader taking his place by the front machine-gun

and the rest of the men sitting on the inward-facing bench seats. The doors were closed by the deputy squad leader, who sat at the rear of the vehicle.

Although the open-topped fighting compartment was uncomfortably exposed both to enemy fire and bad weather, it did have some advantages, including good all-round visibility. Whilst on the move it was usual for one man to scan the skies constantly for enemy aircraft, whilst others kept watch on both sides of the vehicle.

PLATOON ORGANIZATION AND DEPLOYMENT

A panzergrenadier platoon was made up of three squads, with the platoon HQ in a separate vehicle. This was usually the SdKfz 251/10 with a forward-firing 37mm (1.45in) PaK 36 replacing the front-

BELOW: *MOTORISIERT* (MOTORIZED) & *GEPANZERT* (ARMOURED) PANZERGRENADIER BATTALIONS

The armoured battalions were an elite force – in September 1943 the German army, *Waffen-SS* and *Luftwaffe* fielded a total of 226 panzergrenadier battalions, of which only 26 (less than 12 per cent) were equipped with armoured halftracks. The remainder had to rely on trucks, which significantly reduced their cross-country mobility and battlefield capabilities.

1 Panzergrenadier Battalion

UNIT	EQUIPMENT	STRENGTH
Battalion HQ	Motorcycle	7
	Open Blitz Type A	4
	Kübelwagen	5
Heavy Weapons Company		
HQ	Motorcycles	5
	Kübelwagen	4
	Opel Blitz Type A	2
	Maultiers	3
Mortar Ptn	Opel Blitz/80mm mortar	6
Panzerjäger Ptn	Kublewagen	1
	Maultier	1
	Maultier/PaK 40	3
Panzerjäger Ptn	Kublewagen	1
	Maultier	1
	Kfz 81/2.8cm sPzB41	3
Infantry Support Gun Sec	75mm leIG	4
Flamethrower Section	SdKfz 251/16	2
Company (x3)		
HQ	Motorcycle	8
	Kübelwagen	2
Infantry Ptn (x3)	Kublewagen	2
	Opel Blitz Type A	3
Heavy Ptn	Kublewagen	2
	Opel Blitz/MG42 MG	2
Mortar Section	Opel Blitz/80mm mortar	2
	LMG	18
	HMG	4
	80mm mortar	2
	Flamethrower	2

Armoured Infantry Battalion

UNIT	EQUIPMENT	STRENGTH
Battalion HQ & Signals Section	Motorcycle	3
	Kettenkrad	2
	Kübelwagen	3
	SdKfz 251/3	4
	SdKfz 251/8	2
	SdKfz 251/11	2
Heavy Company		
HQ	Motorcycle	2
	Kettenkrad	2
	SdKfz 251/3	2
	SdKfz 11	2
Light Infantry Gun Ptn	SdKfz 251/4	2
	75mm leIG (towed)	4
Infantry Gun Ptn	SdKfz 251/3	1
	SdKfz 251/1	1
	SdKfz 251/9	6
Panzerjäger Ptn 1	SdKfz 251/1	1
	SdKfz 251/4	4
	75mm PaK 40 (towed)	3
Panzerjäger Ptn 2	SdKfz 251/1	1
	SdKfz 251/4/2.8cm sPzB41	3
Company (x3)		
HQ	Motorcycle	2
	Kübelwagen	1
	Kettenkrad	3
	SdKfz 251/3	2
Infantry Ptn (x3)	SdKfz 250/10	1
	SdKfz 251/1	3
Heavy Ptn	SdKfz 250/10	1
HMG Section	SdKfz 251/1	3
Mortar Section	SdKfz 251/2/80mm mortar	2

mounted MG34 or MG42. Initially it was fitted with the standard armoured shield of the towed PaK 36, but this gave the vehicle a distinctive silhouette, making it a prime target for hostile fire. In some units it was common practice to remove the entire shield, or replace it with a much lower version in order to make these vehicles less conspicuous targets. The 37mm (1.45in) was initially intended to give the platoon a mobile anti-tank capability, but the gun's poor armour-piercing performance meant that it was increasingly relegated to firing HE in the close support role. The HQ troop consisted of a platoon commander, usually a junior officer but sometimes a sergeant, a driver, a radio-operator, two messengers and a medical orderly.

On the march, the platoon's vehicles would normally be in close order (5–10m/16ft 5in–32ft 10in apart), moving in column along roads or even a rough line abreast when crossing open country. In combat, however, the formation assumed open order with vehicles separated by 50m (164ft) or more, and would manoeuvre in irregular lines or chequered formations.

If the whole battalion was deployed, the preferred formation was often an arrowhead. In general, SdKfz 251 units rarely exceeded a road speed of 30km/h (18mph), and even under ideal conditions a panzer or panzergrenadier division was not expected to advance more than 20km (12 miles) in a day.

SdKfz 251 drivers generally drove through scattered small arms fire, but would seek cover from enemy artillery or anti-tank guns. The vehicle's machine-gunners might engage targets on the move, sometimes supported by the rest of the squad with rifle and submachine-gun fire from the sides. The SdKfz 251s developed their own version of fire and movement, advancing, stopping and firing to cover other halftracks.

A stationary halftrack made a good firing platform, but was vulnerable to even the lightest anti-tank weapons, and it was not recommended to halt for more than 15–20 seconds in hostile terrain. The normal dismounting procedure was via the rear doors, but in emergencies the squad might also jump over the sides, often whilst the vehicle slowed but was still moving, and the driver then accelerated away to find cover or to take up a firing position.

PULK

One of the most important German formations developed on the Eastern Front was the PULK, a term derived from *panzer und lastkraftwagen* (tanks and trucks). This was a hollow wedge of AFVs containing the panzergrenadiers. The point of the wedge was formed by the most powerful AFVs, the

SdKfz 251/9 Ausf C

■ CREW

■ SPECIFICATIONS

Weight: 8.53 tonnes (8.4 tons)
Length: 5.8m (19ft)
Width: 2.1m (6ft 11in)
Height: 2.07m (6ft 9.5in)
Engine: 74.5kW (100hp) Maybach HL 42 TUKRM 6-cylinder petrol
Speed: 53km/h (33mph)
Range: 300km (186.4 miles)
Armament: 1 x 75mm (2.95in) KwK37 L/24, plus 2 x 7.92mm (0.31in) MG34 or MG42 machine-guns

Büssing-NAG began development of this fire support variant, often referred to as 'Stummel' (stump), in March 1942. Within three months, two prototypes were completed and despatched to the Eastern Front for combat trials, as a result of which an initial 150 vehicles were ordered.

rest making up the sides. The wedge would break through the enemy defences on a narrow front, widening the gap as it passed through. The panzergrenadiers would then fan out to attack remaining areas of resistance from the flanks and rear. If the enemy's weakest point had not been identified, the PULK might advance as a rough rectangle. Once a weak spot was found, the formation could incline left or right, its corner becoming the point of advance.

SdKfz 251/1 Ausf A

■ CREW

■ SPECIFICATIONS

Weight: 9 tonnes (8.85 tons)
Length: 5.8m (19ft)
Width: 2.1m (6ft 11in)
Height: 2.16m (7ft 1in)
Engine: 74.5kW (100hp) Maybach HL 42 TUKRM 6-cylinder petrol
Speed: 53km/h (33mph)
Range: 300km (186.4 miles)
Armament: 2 x 7.92mm (0.31in) MG34 or MG42 machine-guns

The SdKfz 251/1 Ausf A was the first true APC to enter service with any army when the first examples were delivered to 1st Panzer Division in 1939. Although only a few of the Ausf A and Ausf B variants were completed in 1939/40 before production switched to the Ausf C, a handful of surviving vehicles remained in use throughout the war.

SdKfz 251/1 Ausf C

■ CREW

■ SPECIFICATIONS

Weight: 9 tonnes (8.85 tons)
Length: 5.8m (19ft)
Width: 2.1m (6ft 11in)
Height: 2.16m (7ft 1in)
Engine: 74.5kW (100hp) Maybach HL 42 TUKRM 6-cylinder petrol
Speed: 53km/h (33mph)
Range: 300km (186.4 miles)
Armament: 2 x 7.92mm (0.31in) MG34 or MG42 machine-guns

The first examples of the SdKfz 251/1 Ausf A entered service with 1st Panzer Division in 1939. Only small numbers of the Ausf A and Ausf B variants were completed in 1939/40 before production switched to the Ausf C, which incorporated improvements to the stowage arrangements, seating and armour layout. All three types could carry a fully equipped infantry squad with a good degree of protection from small arms fire and shell splinters, although the vehicle's open-topped fighting compartment was vulnerable to indirect fire weapons.

Gepanzert (Armoured) Panzergrenadier Company, 1943

The panzergrenadier company was a key element in the vital 'small change of soldiering', fighting the countless small-scale but deadly actions that decided the outcome of the over all battle.

The company largely depended on its ability to dominate the enemy by traditional 'fire and movement' tactics, with much depending on the ability of its commander, who had to co-ordinate the efforts of a total of almost 200 men. Unlike the commanders of non-mechanized infantry or motorized panzergrenadier companies, he had the huge advantage of being able to deploy a highly mobile force with all its combat elements carried in armoured vehicles.

FIRE SUPPORT

The bulk of the company's combat strength lay in its three standard platoons. The platoon commander's SdKfz 251/10 could give supporting fire with its 37mm PaK 36, whilst the remaining three halftracks generated a high volume of fire from their combined total of six MG34 or MG42.

German tactical doctrine regarded the machine-gun as the infantry squad's primary weapon, with the remainder of the squad dedicated to its care, ammunition supply and protection. Conventional infantry squads had a single machine-gun, but all panzergrenadier squads (motorized and armoured) were equipped with two.

MG34

During the first half of the war, the standard machine-gun was the MG34, which had entered service in the mid-1930s. At the time, it was an entirely new concept – a 'weapons system' in which a single weapon could be fitted with various accessories to allow it to fulfil a variety of roles. This was developed to overcome the problems inherent in the tactical deployment of the various machine-guns of the World War I.

An SS panzergrenadier enjoys a quiet moment with a puppy during Operation 'Citadel'. An MG42 heavy machine-gun lies near to hand.

Armoured Infantry Company: Equipment		
UNIT	EQUIPMENT	STRENGTH
Company HQ	Motorcycle	2
	Kübelwagen	1
	Kettenkrad	3
	SdKfz 251/3	2
Infantry Ptn (x3)	SdKfz 250/10	1
	SdKfz 251/1	3
Heavy Ptn	SdKfz 250/10	1
HMG Section	SdKfz 251/1	3
Mortar Section	SdKfz 251/2/80mm mortar	2

The German army had ended that war with three different weapons: the massive water-cooled Maxim MG08, operating in the sustained fire role; and the hybrid Maxim 08/15 and Madsen guns, both early 'light' machine-guns. In theory, these covered all tactical requirements, but in practice, it was difficult to have the right gun in the right place at the right time.

The MG34 was meant to fulfil all these roles, making it the first general-purpose machine-gun. In its basic form, it was an effective light machine-gun, fed from one of the following:

■ a 250-round ammunition belt, formed by linking five 50-round belts;
■ an assault drum that carried a 50-round belt;
■ a beltless 75-round saddle drum magazine.

In this configuration, it was light enough to be readily carried and fired on the move by a single gunner, although its high rate of fire (800–900rpm) meant that both types of drum magazines were emptied in a few seconds' firing. It was readily adaptable to the sustained fire role when mounted on a sophisticated tripod, the MG Lafette 34, which incorporated high-precision elevation and

traverse controls together with a telescopic sight so that targets could be engaged at ranges out to 3000m (9843ft).

As a final refinement, it could be employed as a light AA weapon on the tall Driebein 34 tripod, using a simple ring sight – pairs of guns were also fitted to the more elaborate Zwillingsockel 36 AA mount.

The MG34 was a very good weapon, precision-made by skilled gunsmiths working to fine tolerances. Unfortunately, it was too well made – the design was poorly suited to mass production

Gepanzert (Armoured) Panzergrenadier Company: Vehicles

Motorcycle: 2
Kübelwagen: 1
Kettenkrad: 3
SdKfz 251/10: 4

SdKfz 251/3: 2
SdKfz 251/1: 12
SdKfz 251/2: 2

The various versions of the SdKfz 251 gave the company a high degree of cross-country mobility and light armoured protection. In addition, it had remarkable firepower for its size – its massed MG42 machineguns could deliver a devastating volume of fire, whilst its integral 8cm (3.15in) mortars provided supporting fire with HE or smoke rounds.

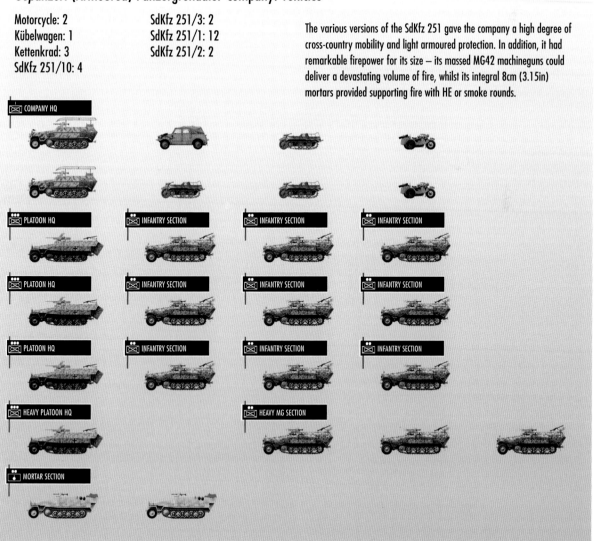

and was very expensive. It also suffered badly in the extreme conditions of the Eastern Front, requiring frequent, careful cleaning to keep it operating in the dust of summer and sub-zero winter temperatures.

MG42

A potential successor, the MG42, had been under development since the late 1930s and the first 17,000 were issued during 1942. (It seems likely that most combat elements of *Das Reich* had re-equipped with the type by the end of the year.) Surprisingly, the designer Ernst Grunow had no experience in weapons design, but attended an army machine-gunner's course to familiarize himself with current weapons whilst seeking input from troops on their battlefield capabilities. He then adapted an existing Mauser-developed operating system and incorporated features suggested by army machine-gunners and lessons learned during the early stages of the war. The new design required far less tooling and was much simpler to build by semi-skilled or even unskilled workers – each gun

could be completed in 75 man hours in contrast to the 150 man hours required for the MG34. Unit costs were also dramatically reduced from 327 RM to 250 RM. A further advantage of the MG 42 was that many components were formed from sturdy stamped sheet metal, which reduced the demands on the limited supplies of high-quality steel.

One of the weapon's most notable features was its high rate of fire of 1200rpm, twice that of the British Vickers machine-gun and the US .30-06 Browning. The MG42 was so intimidating that the US Army made at least one training film in an attempt to dispel the gun's reputation as a 'wonder weapon'. Such a high rate (equating to 20 rounds per second) made a sound frequently likened to ripping cloth, which gave rise to the nicknames 'Hitler's buzzsaw' and 'Hitler's zipper'. (To the Red Army, it was 'the linoleum ripper'.) German troops dubbed it *Hitlersäge* (Hitler's saw), or called it the 'Bonesaw'. Despite the MG42's high rate of fire, German regulations forbade the firing of more than 250 rounds in a single burst and recommended a sustained rate of fire of no more

MG34

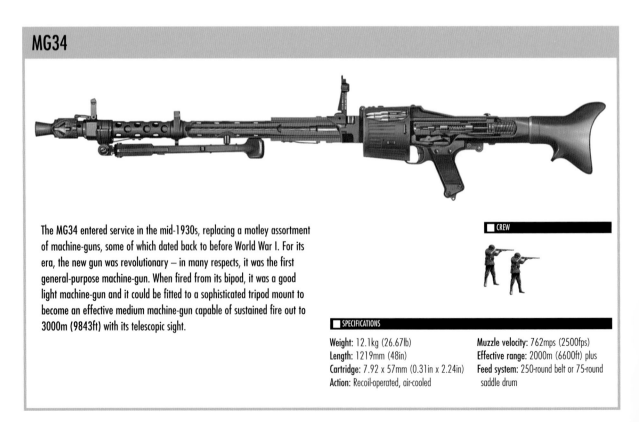

The MG34 entered service in the mid-1930s, replacing a motley assortment of machine-guns, some of which dated back to before World War I. For its era, the new gun was revolutionary – in many respects, it was the first general-purpose machine-gun. When fired from its bipod, it was a good light machine-gun and it could be fitted to a sophisticated tripod mount to become an effective medium machine-gun capable of sustained fire out to 3000m (9843ft) with its telescopic sight.

■ CREW

■ SPECIFICATIONS

Weight: 12.1kg (26.67lb)
Length: 1219mm (48in)
Cartridge: 7.92 x 57mm (0.31in x 2.24in)
Action: Recoil-operated, air-cooled

Muzzle velocity: 762mps (2500fps)
Effective range: 2000m (6600ft) plus
Feed system: 250-round belt or 75-round saddle drum

Gepanzert (Armoured) Panzergrenadier Company: Personnel

Officers: 5
Men: 160

Despite the titles 'Rifle Platoon' and 'Rifle Squad', it should be remembered that German practice was to base infantry tactics on the squad's MG34 or MG42, with the riflemen's prime function being to support their machine-guns. By mid-1943, each panzergrenadier rifle squad fielded two MG42s with a combined rate of fire of approximately 2400rpm.

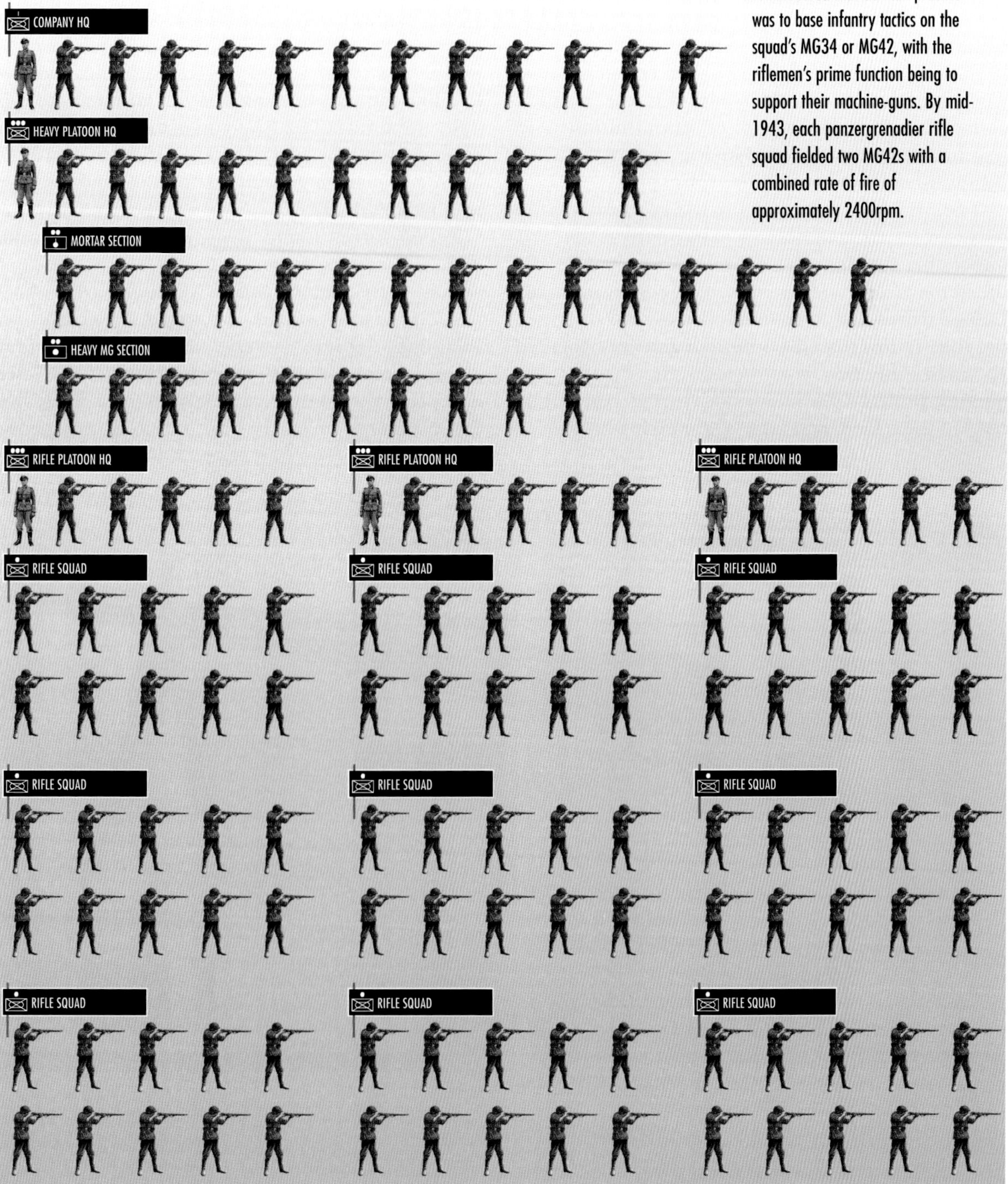

COMPANY HQ

HEAVY PLATOON HQ

MORTAR SECTION

HEAVY MG SECTION

RIFLE PLATOON HQ

RIFLE PLATOON HQ

RIFLE PLATOON HQ

RIFLE SQUAD

RIFLE SQUAD

RIFLE SQUAD

RIFLE SQUAD

RIFLE SQUAD

RIFLE SQUAD

RIFLE SQUAD

RIFLE SQUAD

RIFLE SQUAD

than 300–350 rounds per minute to minimize barrel wear and overheating.

The exceptionally high rate of fire resulted from experiments with earlier weapons which concluded that, as a machine-gunner generally had only brief firing opportunities, it was essential to fire the highest number of rounds possible to increase the likelihood of a scoring a good percentage of hits. The disadvantage of this was that the weapon consumed huge amounts of ammunition and required frequent barrel changes, especially when operating in the sustained fire role.

The MG 42 was designed to use as many MG 34 accessories as possible – although it proved impossible to adapt the complex 75-round saddle drum magazine, and an entirely new version of the Lafette tripod had to be developed for the sustained fire role. In this role, the optimum gun crew totalled six men: the gun commander, the No. 1 (the gunner, preferably a junior NCO), the No. 2 who carried the tripod, and Nos. 3, 4 and 5, who carried

ammunition, spare barrels, entrenching tools and other items. For personal protection the commander, No. 1 and No. 2 were armed with pistols, while the remaining three carried rifles. This could be reduced to just three: the gunner, loader (who also carried the spare barrel) and the commander, although in such cases, it was essential for the rest of the squad to carry extra ammunition to allow the gun to remain in action for more than a few minutes. (The heavy machine-gun section of the company's heavy platoon had three SdKfz 251/1 halftracks, carrying Lafette tripod mounts for their MG34s or MG42s.)

Allied forces found it difficult enough to deal with ordinary German infantry squads each of which were armed with a single MG 42 – panzergrenadier squads equipped with two such weapons apiece were able to generate phenomenally high firepower, which made them exceptionally dangerous opponents. This was especially true on the Eastern Front, where the Red

MG42

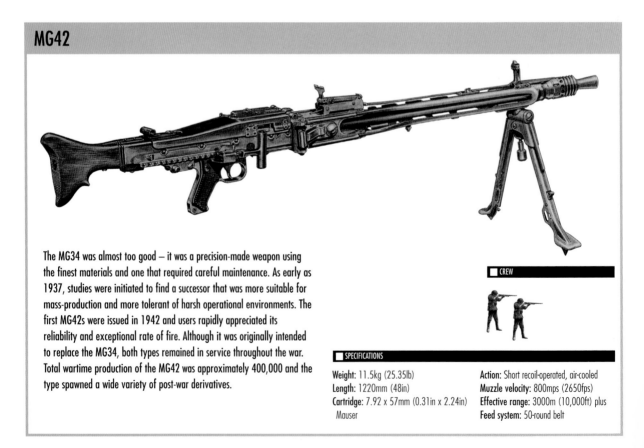

The MG34 was almost too good – it was a precision-made weapon using the finest materials and one that required careful maintenance. As early as 1937, studies were initiated to find a successor that was more suitable for mass-production and more tolerant of harsh operational environments. The first MG42s were issued in 1942 and users rapidly appreciated its reliability and exceptional rate of fire. Although it was originally intended to replace the MG34, both types remained in service throughout the war. Total wartime production of the MG42 was approximately 400,000 and the type spawned a wide variety of post-war derivatives.

CREW

SPECIFICATIONS

Weight: 11.5kg (25.35lb)
Length: 1220mm (48in)
Cartridge: 7.92 x 57mm (0.31in x 2.24in) Mauser

Action: Short recoil-operated, air-cooled
Muzzle velocity: 800mps (2650fps)
Effective range: 3000m (10,000ft) plus
Feed system: 50-round belt

Army's crude infantry tactics often presented massed targets that were literally wiped out by concentrated MG42 fire.

8CM SCHWERE GRANATWERFER 34 (8CM SGRW 34)

The company's heavy platoon gave it a useful integral fire support element. The platoon's mortar section contained two SdKfz 25/2s, each armed with a forward-firing 8cm (3.15in) sGrW 34 in the troop compartment, which was modified to stow up to 66 mortar bombs. The mortar was primarily intended to be fired from the vehicle, although a standard base plate was carried, allowing it to be dismounted and ground-fired from positions which were otherwise inaccessible.

Initially the sGrW 34 was restricted to HE and smoke ammunition, but a wide range of 8cm (3.15in) ammunition was developed throughout the war, including an unusual 'bouncing bomb'. This was the Wurfgranate 38, which had a small extra charge in the nose that detonated on impact, blowing the bomb back into the air. This initiated a short delay fuse, which exploded the bomb several metres above the ground – such air bursts produced far more lethal fragmentation than conventional mortar ammunition.

INFANTRY ANTI-TANK WEAPONS

This was the area of greatest weakness in German equipment in mid-1943. The anti-tank rifles of the early war years were totally obsolete and the first Panzerfausts and Panzerschrecks would not appear until after Operation 'Citadel'. In the meantime, even elite formations such as *Das Reich* had to rely on an assortment of weapons of varying effectiveness, all of which demanded immense courage and skill to use with any chance of success.

Amongst the most important of these were:

■ **Geballte Ladung.** The Model 24 stick grenade was too light to cause anything other than superficial damage to AFVs and fortifications, but it was frequently used as the basis for an effective improvised demolition charge – the heads of six grenades were removed and wired around a central stick grenade. This was termed *Geballte Ladung* (literally 'baled charge' or 'concentrated charge'). Although the device was heavy and awkward, with a very short throwing range, it was quite capable of

wrecking the engine of any tank if it could be lobbed onto the rear decking. It was equally effective against all but the strongest field defences and was used throughout the war.

■ **Gross Gewehr Panzergranate 40 (GPzgr 40).** This spin-stabilized HEAT rifle grenade was effectively an enlarged version of the Gewehr Panzergranate 30, which was fired from the Kar98k by an unusual wooden-bulleted blank round. Its effective range was roughly 91m (299ft) and it could penetrate up to 90mm (3.54in) of armour.

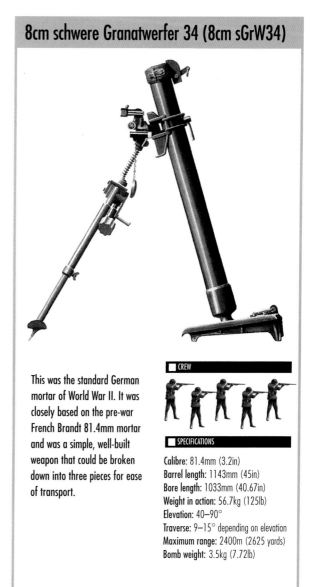

8cm schwere Granatwerfer 34 (8cm sGrW34)

This was the standard German mortar of World War II. It was closely based on the pre-war French Brandt 81.4mm mortar and was a simple, well-built weapon that could be broken down into three pieces for ease of transport.

■ CREW

■ SPECIFICATIONS

Calibre: 81.4mm (3.2in)
Barrel length: 1143mm (45in)
Bore length: 1033mm (40.67in)
Weight in action: 56.7kg (125lb)
Elevation: 40–90°
Traverse: 9–15° depending on elevation
Maximum range: 2400m (2625 yards)
Bomb weight: 3.5kg (7.72lb)

■ **3kg Haft-Hohlladungen (Haft-Hl 3).** This device was first issued in November 1942 and was intended as a dual-purpose demolition/anti-tank weapon. Whilst it could be thrown at close range, its size and weight made it more practical to place it on enemy AFVs. The three sets of powerful magnets held the HEAT warhead at the optimum stand-off distance to ensure maximum armour piercing performance – 140mm (5.5in). When used as a demolition weapon, it could penetrate up to 50cm (20in) of concrete. Almost 554,000 were produced between 1942 and May 1944, when the type was officially replaced by the Panzerfaust. However, it was ordered that existing stocks should be used up and the type remained in service until the end of the war.

■ **Tellermine 35 (T.Mi.35 (S)).** This was the most common German anti-tank mine – it had a sheet steel case and a slightly convex pressure plate on the top surface with a central fuse well. Two secondary fuse wells for anti-handling devices were located on the side and bottom of the mine. Pressure of 180kg (400lb) on the centre of the mine or 90kg (200lb)

An 8cm (3.2in) mortar team poses for the camera. All the men are wearing the SS-specific summer pattern camouflage smock and helmet cover. The high angle of the mortar tube indicates the closeness of the target.

on the edge of the mine deformed the pressure plate that compressed a spring. This sheared a retaining pin holding the striker, which was driven into a percussion cap to ignite the detonator followed by the booster and main charges. The mine's total weight was 9.1kg (20lb), 5.5kg (12.1lb) of which was explosive (usually TNT), a charge sufficient to break the tracks and damage the suspension of virtually all wartime AFVs.

Although not generally thought of as an infantry anti-tank weapon, the Tellermine was commonly used in this role, with each squad carrying several mines. In action, they could be thrown from cover into the path of oncoming tanks. Alternatively, they were sometimes fitted with time fuses and either thrown onto the tanks' engine decks or (especially against T-34s and KV-1s) jammed under the overhang at the rear of the turret. The latter method was highly effective, invariably knocking out the tank and usually blowing its turret off.

■ **Demolition charges.** These were time-fused containers of explosives (usually TNT) issued in various sizes, most commonly 1kg (2.2lb), 3kg (6.6lb) and 10kg (22lb). In the anti-tank role, they were generally used in various combinations to make up satchel charges that could be thrown beneath

enemy tanks or onto their engine decks. One device which required even greater skill and courage to use was the *Doppelladung* (double charge), which comprised two 1kg (2.2lb) charges linked by short lengths of wire. These were intended to be thrown over a tank's gun barrel.

Karabiner 98 Kurz (Kar98k)

For much of the inter-war period, the German army was equipped with the Karabiner 98b, a slightly modified version of the unwieldy Gewehr 98 rifle of World War I. This was replaced by the Kar98k in 1935, which was essentially a modernized, shortened version of the Gewehr 98. The new weapon was a robust and accurate bolt-action rifle that remained in service until the end of the war. An estimated 14,000,000 of these rifles were completed between 1935 and 1945.

SPECIFICATIONS

Weight: 3.7–4.1kg (8.2–9lb)
Length: 1110mm (43.7in)
Cartridge: 7.92 x 57mm IS
Action: Bolt-action
Muzzle velocity: 760m/s (2493 ft/s)

Effective range: 500m (1640ft)
(with iron sights) / 800+m (2625ft)
(with optics)
Feed system: 5-round stripper clip, internal magazine

MP40

SPECIFICATIONS

Weight: 4kg (8.82lb)
Length: 833mm (32.8in) stock extended /
630mm (24.8in) stock folded
Cartridge: 9 x19mm (0.35 x 0.75in)
Parabellum

Action: Straight blowback, open bolt
Muzzle velocity: 380m/s (1247ft/s)
Effective range: 100m (328ft)
Feed system: 32-round detachable box
magazine

Until the late 1930s, virtually all submachine-guns were heavy wooden-stocked weapons made using traditional gunsmiths' skills. At the time of its introduction in 1938, the MP38 represented a dramatic break with those traditions – it was virtually an all-metal weapon with a folding stock and a Bakelite synthetic pistol grip. Early combat experience showed that there was a real risk of the gun firing if was dropped, and the safety arrangements were significantly improved in the next model, the very similar MP40, which also simplified for ease of production. Many MP38s were also modified to improve their safety and were designated MP38/40s. The series remained in service throughout the war, with a total production of almost 1,000,000 of all types.

SdKfz 251/10

CREW

SPECIFICATIONS

Weight: 8 tonnes (7.87 tons)
Length: 5.8m (19ft)
Width: 2.1m (6ft 11in)
Height: 2.07m (6ft 9.5in)
Engine: 74.5kW (100hp) Maybach HL
42 TUKRM 6-cylinder petrol
Speed: 53km/h (33mph)
Range: 300km (186.4 miles)
Armament: 1 x 37mm (1.45in)
PaK35/36 L/45, plus 1 x 7.92mm
(0.31in) MG34 or MG42 machine-gun

This version was issued to platoon commanders as the *Zugführerwagen*. The 37mm (1.45in) PaK35/36 eventually became as much an infantry support weapon as an anti-tank gun. Initially, the standard gun shield was fitted, but the height of this made the vehicle dangerously conspicuous and it was generally replaced with smaller shields of varying designs.

Gepanzert (Armoured) Panzergrenadier Heavy Weapons Company, 1943

The heavy weapons company provided its parent battalion with an invaluable anti-tank capability and a source of almost instant fire support.

The bulk of the company's assorted weapons were self-propelled, which greatly enhanced both their mobility and their battlefield survivability.

SUPPORT WEAPONS

The experience of World War I showed that conventional artillery could rarely respond quickly enough to help frontline infantry deal with isolated strong points that were holding up their advance. By 1918, many armies had light artillery, or 'infantry guns', as an integral part of their infantry units. Whilst the French developed a new 37mm (1.45in) gun specifically for this role, the Germans favoured modified 77mm (3in) field guns that fired a much more destructive shell.

The value of these guns led the post-war German army to issue a requirement for a purpose-built 75mm (2.95in) infantry gun in the mid-1920s. In 1927, an order was placed with Rheinmetall for a design featuring an unusual 'shotgun' breech action. After firing, the barrel slid back over the breech block, with the muzzle pointed toward the ground and the breech end pointing upwards. The crew then slid the next shell into the barrel, which automatically closed and locked the breech, ready to fire again.

The gun's compact design made it easy to conceal, even in relatively open terrain. Its light weight was another feature greatly appreciated by its crews, who frequently had to manhandle their pieces over quite long distances to take up suitable firing positions. This was especially true of those in conventional infantry divisions who had to rely on horse teams, which were too vulnerable to risk in the front line.

In contrast, *Das Reich*'s infantry gun crews had a relatively easy life – their guns were towed by SdKfz 251/4s, whose armour frequently allowed them to move right up to the selected firing positions. They could also carry a good supply of the small 75mm (2.95in) shells, which included a proportion of HEAT rounds for emergency anti-tank defence.

These towed weapons were supplemented by the company's six SdKfz 251/9 self-propelled 7.5cm KwK 37 L/24 guns. The guns were surplus weapons that were originally produced for early Panzer IVs, but which were given a new lease of

Heavy Weapons Company: Equipment

UNIT	EQUIPMENT	STRENGTH
Company HQ	Motorcycle	5
	Kübelwagen	4
	Opel Blitz Type A	2
	Maultier	3
Mortar Ptn	Opel Blitz/80mm mortar	6
Panzerjäger Ptn	Kublewagen	1
	Maultier	1
	Maultier/PaK 40	3
Panzerjäger Ptn	Kublewagen	1
	Maultier	1
	Kfz 81/2.8cm sPzB41	3
Infantry Support Gun Section	75mm leIG	4
Flamethrower Section	SdKfz 251/16	2

An anti-tank crew from the *Das Reich* division load a heavily-camouflaged PaK 40 with an armour-piercing round in anticipation of a Soviet armoured assault.

life in this configuration, proving to be highly successful close support weapons. Each SdKfz 251/9 stowed 52 rounds of 75mm (2.95in) ammunition, which could include HE, HEAT, canister and smoke rounds.

FLAMETHROWERS

The SdKfz 251/16 was a highly specialized vehicle equipped with two side-mounted flamethrowers that were fed from internal tanks holding a total of 700l (154 gallons), sufficient for 80 two-second bursts of flame. Each of these had a total traverse of 160° and their operators were protected by small armoured shields. The initial production batches,

Private, Panzergrenadier Company

The distinctive Waffen-SS *camouflage smock and helmet cover originated in the late 1930s with experiments carried out by two officers on the staff of the* Deutschland *regiment,* SS-Obersturmführer *Ecke and* SS-Hauptsturmführer *Dr. Ing.* Wim Brandt. *This example is one of a number of 'spring/summer' designs with a predominately green pattern. The private is armed with the robust and accurate Kar98k bolt-action rifle, which remained in service until the end of the war. An estimated 14,000,000 of these rifles were completed between 1935 and 1945.*

which entered service in January 1943, were also fitted with a portable flamethrower derived from the standard man-pack Flammenwerfer 42. This was connected to the vehicle's flame fuel tanks by a 9m (33ft) hose and was intended for use against targets that could not be reached by the larger flame guns. Whilst this was a theoretically useful capability, it seems to have been little used in practice, and the portable flamethrower was omitted from later production vehicles.

The flame fuel was fed from the storage tanks by a pump that was powered by a small auxiliary petrol engine, generating sufficient pressure to allow shots of up to 36.5m (120m) from the main flame guns. The effective range of the portable flamethrower was approximately 27.5m (90ft).

In addition to its main armament, the SdKfz 251/16 carried two MG34 or MG42 machine-guns and was manned by a crew of four – driver, radio operator/ machine-gunner and two flamethrower operators.

FLAMETHROWER TACTICS

The SdKfz 251/16 was a terrifying weapon, but one that required careful tactical handling due to its thin armour and the fire risk posed by its 700l (154 gallons) of flame fuel. The expectation was that it would use its machine-guns for suppressive fire against personnel and other 'soft' targets at ranges up to 400m (1312ft), only firing the flame guns against personnel and static targets within 36m (118ft). (Not unreasonably, the type's tactical notes assumed that even if the flame did not destroy enemy troops, it should at least force them out of cover.) Combat experience showed that flame attacks were particularly effective in mopping up operations and in destroying enemy personnel in field fortifications.

The SdKfz 251/16 halftracks were normally deployed in full platoon strength and always in close cooperation with mounted panzergrenadier units in the attack. However, it was recognized that there were circumstances (such as urban combat and assaults on permanent fortifications) in which they needed to be employed singly, under the command of mounted panzergrenadier platoons. It was therefore standard procedure for the flamethrower vehicles to remain with the APCs when the panzergrenadiers dismounted.

Heavy Weapons Company

Motorcycle: 5	75mm leIG: 4	Opel truck with 8cm mortar: 6
Kübelwagen: 5	Maultier: 4	Maultier towing 3 x PaK 40: 3
Opel truck: 2	SdKfz 251/16: 2	Kfz 81 towing 3 x 2.8cm sPzB41: 3

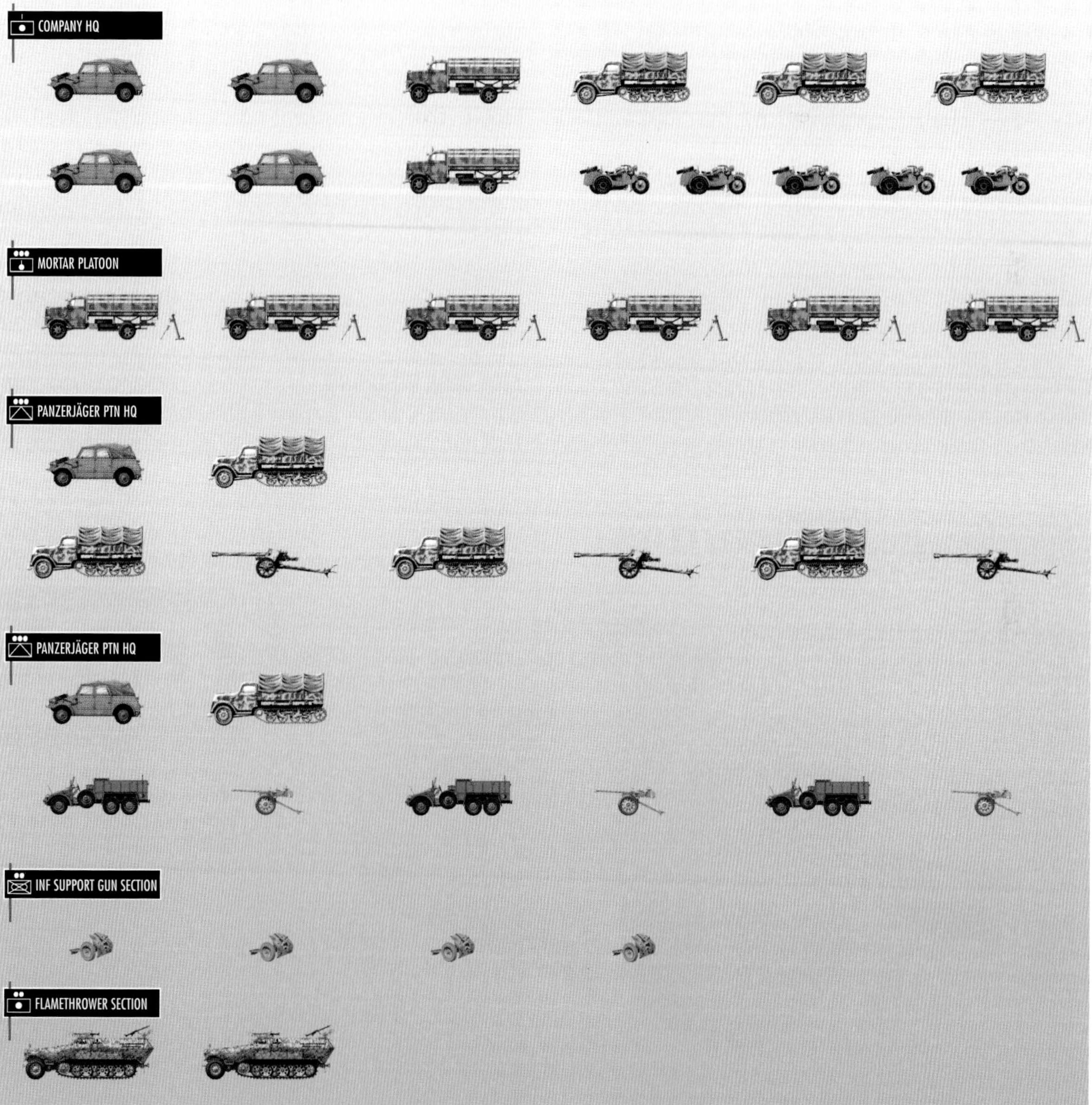

COMPANY HQ

MORTAR PLATOON

PANZERJÄGER PTN HQ

PANZERJÄGER PTN HQ

INF SUPPORT GUN SECTION

FLAMETHROWER SECTION

Tactical instructions emphasized that these highly specialized vehicles should not be used as substitutes for panzers or assault guns, as 'point' vehicles on the march or in action, or as independent patrol vehicles.

In the attack, the flamethrower vehicles advanced in open order behind the mounted panzergrenadier units, which would lay down suppressive machine-gun fire from the start of the action, enabling the flamethrower platoon to break into the enemy position. If the opposition remained under cover, it would be burnt out. (Vehicle crews were cautioned that bursts of fire from the flamethrowers should be directed only against those targets which were definitely within range; combat experience had shown that indiscriminate bursts of flame fired at extreme range merely wasted fuel and obscured vision.)

Flame gunners were instructed to direct the flame against the bottom of the target first and then work up, to ensure the destruction of any enemy infantry anti-tank teams. However, it was recognized that the type of target and the progress of the attack would determine whether fire should be opened whilst on the move or at the halt.

It was recommended that large targets should first be sprayed with flame fuel and then ignited by a burst of fire. This was found to be especially effective when attacking dugouts, trenches, pillbox entrances or firing ports and wooden buildings. (In some cases, this would cause an explosion, due to a rapid build-up of inflammable fumes inside the target structure. Even if no explosion occurred, the ignition would inflict casualties because it sucked the oxygen from the room or bunker.)

Although flamethrowers had a very limited range, the effects of friendly fire from flame weapons could be truly horrifying and crews were banned from opening fire in thick fog without specific orders.

ANTI-TANK WEAPONS

The unit's most potent anti-tank weapon was the PaK 40 platoon (described in the section on the

7.5cm leichtes Infanteriegeschütz 18 (7.5cm le.IG 18)

CREW

SPECIFICATIONS

Weight in action: 0.4 tonnes (0.39 tons)
Calibre: 75mm (2.95in)
Muzzle velocity: 210m/sec (689ft/sec)
Range (HE): 3550m (11,650ft)
Ammunition: HE, HEAT
Shell weight (HE): 5.5kg (12.12lb)

This diminutive infantry gun was adopted by the *Reichswehr* in 1932 and remained in service until the end of the war. Early examples were horse-drawn or man-handled and were fitted with wooden-spoked wheels, whilst a proportion of later models had metal disc wheels with pneumatic tyres for use by motorized units. During the second half of the war, a HEAT shell capable of penetrating 45–55mm (1.7–2.2in) of armour was produced to give the gun a degree of anti-tank capability.

Panzerjäger battalion, pages 107–110). This was supplemented by a platoon of the unique cone-bore 2.8cm schwere Panzerbüchse 41 (2.8cm sPzB41). The cone-bore idea was originally patented in 1903 with the aim of producing an ultra-high velocity hunting rifle, but was transformed to a practical system by the work of a German engineer, Hermann Gerlich, who produced an experimental 7mm (0.26in) anti-tank rifle in the mid-1930s, with an astonishingly high muzzle velocity of 1800m/sec (5900ft/sec).

This provided the basis for Mauser-Werke AG's development of a 28/20mm (1.1/0.7in) anti-tank gun initially designated Gerät 231 or MK8202. In June/July 1940, an initial batch of these were sent for service trials which resulted in some modifications. Manufacture of the new gun began in 1941 and approximately 2800 were completed by 1943, when production was halted due to increasing shortages of the tungsten required for its ammunition.

GUN OR RIFLE?

Although the sPzB 41 was officially classified as a heavy anti-tank rifle, its construction was much more typical of an anti-tank gun. Like the latter, it had a recoil mechanism, carriage and shield. The heart of the design was the barrel with its unique cone-shaped bore, which reduced from 28mm (1.1in) at the chamber to only 20mm (0.7in) at the muzzle.

The armour-piercing round was an entirely new design, described as armour-piercing composite non-rigid (APCNR) and officially designated 2.8cm Pzgr.41. It had a tungsten carbide core within a soft steel casing and was fitted with a magnesium alloy ballistic cap. The core was 40mm (1.5in) long and 10.9mm (0.43in) in diameter, with a tungsten content of about 9.1 per cent. The casing carried two external flanges and, on firing, these were squeezed down as the round moved down the barrel, decreasing the surface area and increasing the muzzle velocity to 1430m/sec (4690ft/sec).

At short ranges, the result was deadly for such a light weapon – at 100m (328ft) the round penetrated at least 52mm (2in) of armour sloped at 60°, whilst even at 500m (1640ft), penetration was 40mm (1.5in). As the gun was essentially a close-range weapon, it was fitted with open sights for use up to 500m (1640ft), although the PaK 36's ZF 1x11 telescopic sight could also be fitted.

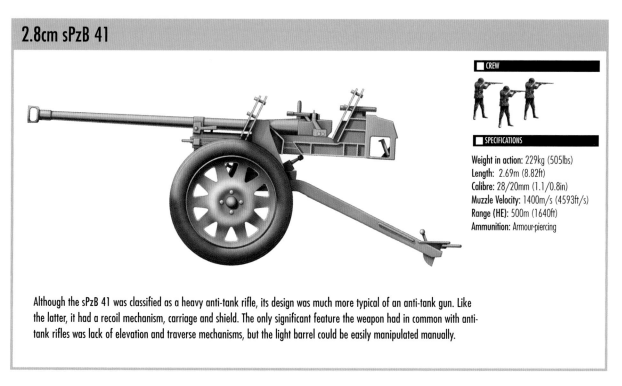

2.8cm sPzB 41

CREW

SPECIFICATIONS

Weight in action: 229kg (505lbs)
Length: 2.69m (8.82ft)
Calibre: 28/20mm (1.1/0.8in)
Muzzle Velocity: 1400m/s (4593ft/s)
Range (HE): 500m (1640ft)
Ammunition: Armour-piercing

Although the sPzB 41 was classified as a heavy anti-tank rifle, its design was much more typical of an anti-tank gun. Like the latter, it had a recoil mechanism, carriage and shield. The only significant feature the weapon had in common with anti-tank rifles was lack of elevation and traverse mechanisms, but the light barrel could be easily manipulated manually.

Self-propelled Heavy Infantry Gun Company

The 15cm sIG 33 (schweres Infanterie Geschütz 33) was the standard German heavy infantry gun from the late 1920s until the end of the war, but was always plagued by problems caused by its excessive weight.

The success of modified 77mm (3in) field guns in the infantry support role during World War I led the Germans to investigate the possible use of larger calibre weapons for the same purpose. The restrictions imposed on the German army's conventional artillery by the Treaty of Versailles also contributed to a readiness to accept unorthodox solutions in the search for greater firepower.

SP Heavy Infantry Gun Company

UNIT	EQUIPMENT	STRENGTH
SP Heavy Infantry Gun Coy	Kettenkrad	5
	Kübelwagen	4
	Maultiers	9
	SdKfz 251/3	1
	SdKfz 251/1	3
	SdKfz 251/11	1
	150mm sIG PzKpwg 38(t)	6

THE sIG 33

The sIG 33 was the largest calibre weapon ever designed as an infantry gun – development began in 1927, but the design was not approved for service until 1933 and it was 1936 before the first units received their guns. Although it was officially described as a gun, the sIG 33 was in fact a howitzer with a short (L/11) barrel, firing a 38kg (84lb) HE shell to an effective range of only 4700m (15,420ft). This restricted range was adequate for its intended close-support role, but the design suffered from excessive weight.

In the late 1930s, the gun was redesigned to incorporate light alloys which cut the total weight by approximately 150kg (330lb), and it seems likely that a small number of these modified weapons were completed in 1938/39. The outbreak of war forced a reversion to the original design as the *Luftwaffe* received priority for the limited supplies of light alloys.

Combat experience in Poland showed just how limited the sIG 33's mobility really was – those equipping motorized and panzer units had difficulty in keeping up with the speed of the advance and it was clear that some form of self-propelled version was urgently required. The first such vehicle was the 15cm sIG 33 (Sf) auf Panzerkampfwagen I Ausf B – a total of 38 Panzer Is were converted in the winter of 1939/40, the last of which remained in service until 1943. The tank's turret and superstructure were removed and a platform was built out over the tracks to take a complete gun

(including its wheeled carriage), protected by a three-sided, open-topped shield. The whole conversion was extremely crude and only just worked – the vehicle's overall height was 2.8m (9ft 2in) and its high centre of gravity made it unstable, limiting its off-road mobility. The chassis was also badly overloaded, resulting in frequent breakdowns, and it was obvious that the type could be regarded as a stop-gap measure only, to give time for the development of a replacement.

THE 'GRILLE'

A more practical conversion using an enlarged Panzer II hull was designed in 1940/41, but only 12 were produced for combat trials in North Africa. The limitations of this type caused attention to be turned to a version based on the versatile Panzer 38(t). Development of this variant began in 1942 and proved to be relatively straightforward. An initial order was placed for 200 examples of what was generally referred to as the 'Grille' (Cricket) although the vehicle appears in some sources as the Bison. It had been intended to use the new Panzer 38(t) Ausf M hull with a centrally mounted engine that was under development by BMM (Böhmisch-Mährische Maschinenfabrik) at Prague, but delays with this version forced the use of the then current Ausf H hull with its conventional rear-mounted engine.

A total of 91 vehicles (including a single prototype) were built between February and April

1943, using both newly completed and repaired hulls. The tank's turret and forward superstructure were removed to create space for an enlarged open-topped fighting compartment for the sIG 33 on a limited traverse mounting. (An MG34 or MG42 was also stowed in the fighting compartment to give a degree of self-defence capability against air attacks and infantry anti-tank teams.) The side armour of the fighting compartment was extended to enclose the engine decks in order to give sufficient space for ammunition stowage, and the vehicle was officially designated 15cm Schweres Infanteriegeschütz 33 (Sf) auf Panzerkampfwagen 38(t) Ausf. H (SdKfz 138/1).

The type was a vast improvement on the earlier self-propelled sIG 33 variants, with a far better

SP Heavy Infantry Gun Company

Kettenkrad: 5
Kübelwagen: 4
Maultier: 9

SdKfz 251/3: 1
SdKfz 251/1: 3
SdKfz 251/11: 1
sIG PzKpfw 38(t): 6

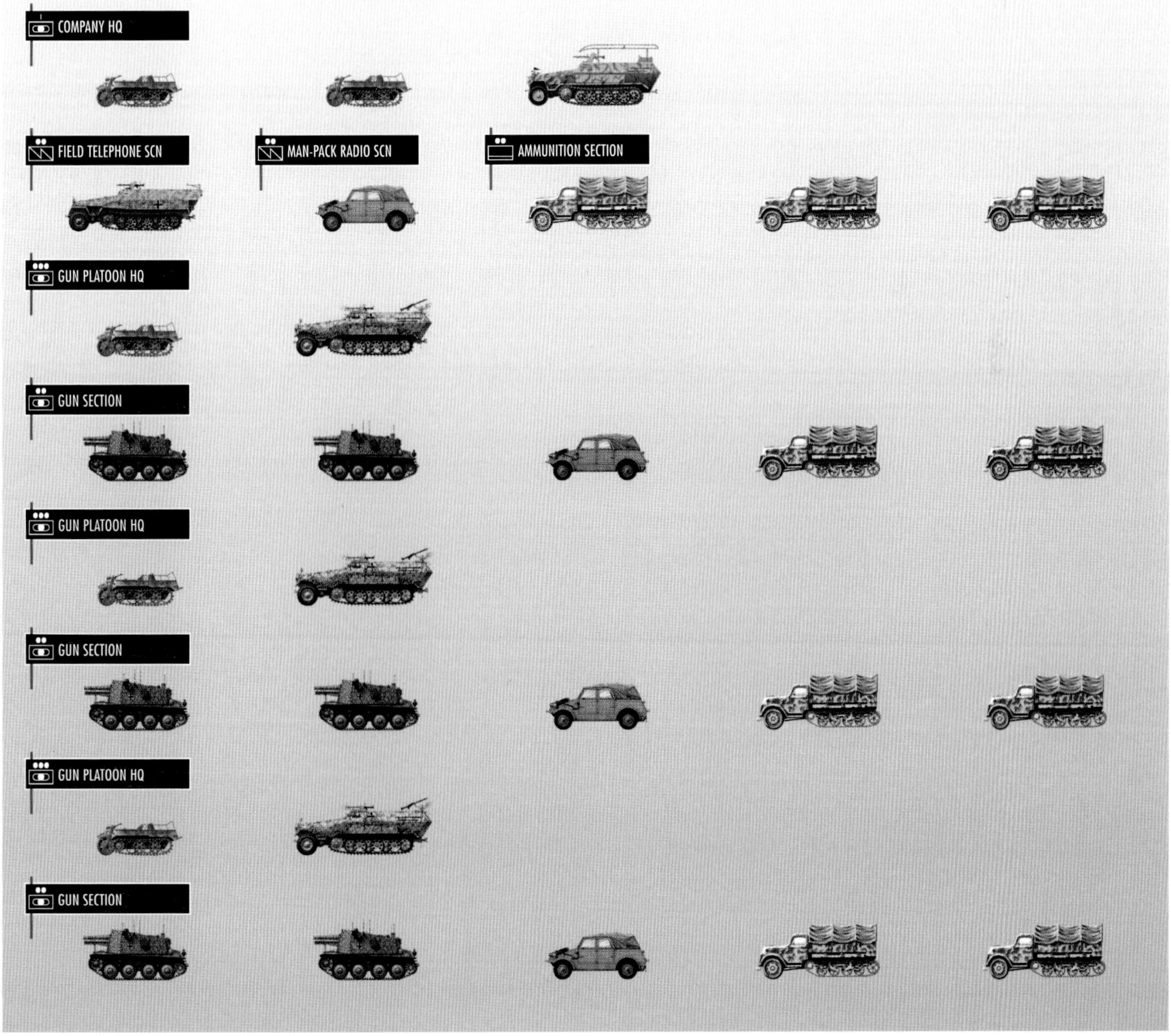

COMPANY HQ

FIELD TELEPHONE SCN   MAN-PACK RADIO SCN   AMMUNITION SECTION

GUN PLATOON HQ

GUN SECTION

GUN PLATOON HQ

GUN SECTION

GUN PLATOON HQ

GUN SECTION

automotive performance and greater overall reliability. Although it still required careful tactical handling in view of its thinly armoured open-topped fighting compartment (with a maximum thickness of only 25mm/0.98in), the armour layout was well designed, including a sprung flap that moved to close the gap in the front plate as the gun was elevated. Main armament ammunition stowage was limited to a maximum of 18 rounds, but this was a relatively minor drawback.

Despite generally favourable assessments by the vehicles' crews, there were some criticisms, especially of the position of the main armament. Mounting this heavy weapon so far forward placed a considerable load on the front suspension units and caused a significant degree of nose-heaviness. These factors impaired the vehicle's cross-country

In the heat of the summer, an outpost of Das Reich panzergrenadiers keeps watch amidst the unharvested crop.

performance and also imposed a heavy workload on maintenance crews, who frequently had to replace damaged front road wheels and suspension components. The problem had been anticipated and production was switched to the Ausf M as soon as development of the type was completed in April 1943 under the official designation 15cm Schweres Infanteriegeschütz 33/1 auf Selbstfahrlafette 38(t) (Sf) Ausf M (SdKfz 138/1). The centrally mounted engine allowed the fighting compartment to be repositioned at the rear, which improved the weight distribution and the vehicle's off-road performance. It proved to be impossible to increase the on-board ammunition stowage, and an ammunition carrier variant was accordingly designed with racks for shells and cartridges replacing the main armament. A total of 282 vehicles were completed between April 1943 and September 1944, as well as a further 120 ammunition carriers.

sIG 33 PzKpfw 38(t) 'Grille'

■ **SPECIFICATIONS**

Weight: 12.19 tonnes (12 tons)
Length: 4.95m (16ft 3in)
Width: 2.15m (7ft 1in)
Height: 2.47m (8ft 1in)
Engine: 111kW (150hp) Praga AC
　　6-cylinder petrol
Speed: 42km/h (26mph)
Range: 190km (118 miles)
Armament: 1 x 150mm (5.9in) Sig
　　33/2 infantry gun, plus 1 x 7.92mm
　　(0.31in) MG34 or MG42 machine-gun

This vehicle was based on the Panzer 38(t) Ausf H, and retained its rear-mounted engine. The tank's turret was removed and replaced by a low-slung superstructure and open-topped fighting compartment for the 15cm schweres Infanteriegeschütz 33 (heavy infantry gun). A total of 91 vehicles (including a single prototype) were produced in the BMM (former âKD Praga) factory in Prague from February to April 1943. The official designation was 15cm Schweres Infanteriegeschütz 33 (Sf) auf Panzerkampfwagen 38(t) Ausf H (SdKfz 138/1).

Self-propelled Flak Company

The company's combat element comprised 12 SdKfz 10/4 or SdKfz 10/5 self-propelled 20mm AA guns. These were heavily modified versions of the standard Sd Kfz 10 light halftrack – the SdKfz 10/4 carried the 2cm FlaK 30 mount on a special platform with fold-down side and rear panels.

This platform was specifically designed for the FlaK 30 mount and could not readily accept a FlaK 38 mount or vice versa. The basic SdKfz 10 was enlarged to accommodate the gun mount and was fitted with four folding seats on the platform for the gun's crew. Some examples were fitted with a gun shield to provide a limited degree of protection for their otherwise very exposed crews. The ready-use ammunition bins attached to the side and rear panels (four on each side and two at the rear) each held a single 20-round clip. To supplement this very limited stowage, it was normal practice to tow an ammunition trailer, the Sonderanhänger 51 (Sd Ah

51, 'special single-axle trailer 51'), which contained a further 640 rounds together with the gun's sights and its rangefinder.

Early production vehicles were fitted with removable loading ramps, cable rollers to act as pulleys, and a reinforced tailgate to allow a FlaK 30, mounted on an Sd Ah 52 trailer, to be quickly dismounted. From 1940 the vehicles were also fitted with rifle racks over the front mudguards, and from 1942 these were modified with the addition of sheet metal covers to give protection from the weather. FlaK 38s were mounted on 10/4s from 1941, although initially the platform was not widened to accommodate the larger gun

Self-propelled Flak Company

Motorcycle: 12
Kübelwagen: 7
Maultier: 3

SdKfz 10: 1
Kfz 17 radio car: 5
SdKfz 10/5: 12

Each self-propelled flak company gave its parent regiment a powerful and highly mobile source of AA fire, although the increasing numbers of well-armoured 'Shturmoviks' posed problems until more armour-piercing ammunition became available. In addition to its primary role, the company was often in deployed against ground targets, in particular to provide fire support for infantry attacks.

SP Flak Company: Equipment

UNIT	EQUIPMENT	STRENGTH
Self-propelled Flak Company	Motorcycle	12
	Kübelwagen	7
	Maultier	3
	Kfz 17 radio car	5
	SdKfz 10	1
	SdKfz 10/5	
	20mm cannon	12

vehicles received similar armour protection from 1943 onwards.)

Although neither variant could compete with the later types of self-propelled Flakvierling in terms of sheer firepower, the smaller and lighter Sd Kfz 10 remained in production until well into 1944. This was partly because it was able to use weaker bridges and had superior cross-country mobility. Such attributes were particularly valuable on the Eastern Front, where there were few metalled roads and even fewer bridges constructed to Western European standards.

Production of the Sd Kfz 10/4 began in 1939 with deliveries to both the army and *Luftwaffe*, although the exact distribution will never be known, as details were often omitted from the various production reports. (An unknown, but probably substantial number were later issued to *Waffen-SS* formations, including *Das Reich*.) Adler built 1054 between 1939 and February 1943, although some of these were completed as 10/5s from 1942 onwards. Mechanische Werke in Cottbus was awarded two contracts for a total of 975 Sd Kfz 10/5s to be delivered in 1943/44.

mounting, a decision that left very little space for the crew.

The designation SdKfz 10/5 was retrospectively applied to those vehicles fitted with the 2cm FlaK 38, but it was not until 1942 that all those coming off the production lines were fitted with enlarged platforms. The *Luftwaffe* ordered 293 sets of 8mm (0.3in) armour plate (*Behelfspanzerung*) for its vehicles in 1943. These plates covered the radiator, windscreen and both sides of the driver's compartment and were fitted to both versions. (It seems likely that many army and *Waffen-SS*

2cm Flakvierling 38 (Sf) auf Fgst Zgkw 8t Sd.Kfz.7/1

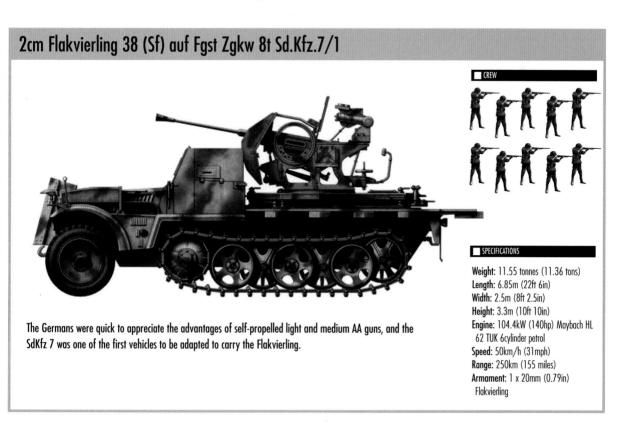

The Germans were quick to appreciate the advantages of self-propelled light and medium AA guns, and the SdKfz 7 was one of the first vehicles to be adapted to carry the Flakvierling.

CREW

SPECIFICATIONS
Weight: 11.55 tonnes (11.36 tons)
Length: 6.85m (22ft 6in)
Width: 2.5m (8ft 2.5in)
Height: 3.3m (10ft 10in)
Engine: 104.4kW (140hp) Maybach HL 62 TUK 6cylinder petrol
Speed: 50km/h (31mph)
Range: 250km (155 miles)
Armament: 1 x 20mm (0.79in) Flakvierling

2/2nd SS Panzer Regiment, 1943

The panzer regiment was arguably the most powerful of all the division's units, but was handicapped throughout Operation 'Citadel' by the absence of its 1st Battalion. which was in Germany re-equipping with the new Panther medium tank.

Despite this, the regiment performed creditably throughout almost two weeks of intensive combat though it was heavily dependent on Panzer IIIs, which were rapidly coming to the end of their effectiveness as battle tanks.

Although much of its equipment was of mediocre quality, the regiment was able to compensate for this by its exceptional battlefield skills. These were summarized in the following tactical notes which were based on combat experience on the Eastern Front:

'The Panzer Regiment is, by reason of its firepower, protection and mobility the main fighting power of the Division. Its strength lies in unexpected, concentrated and determined attack; aggressive leadership and daring operations. Combat in Russia has shown once again that in action against the Communists, it is not so much the kind or number of our tanks but the spirit and skill of the tank soldiers that count. It is only due to these factors that German tanks are victorious in Russia. This exemplary combat spirit will, however, count for little if our tanks are not led and employed

by fully competent officers. Superior tactical leadership in battle is essential to minimize casualties.

The purpose of this volume is to collect the experiences of the veteran frontline combat leaders of our Regiments in action, and pass it on in simple and understandable form to our junior officers.

1. Before any attack acquaint yourself with the ground. Use information provided by other units or by the map and share this information with your subordinates. Correct assessment of the terrain can make the difference between victory and defeat.
2. Even under the most extreme circumstances, no attack is so urgent that you do not have time to put subordinates in the picture about the tactical situation, the mission and anything else which may affect the coming action. Losses due to over-hasty action are your responsibility and jeopardize the success of the mission.
3. Only careful combat reconnaissance can protect

ORBAT: SS *Das Reich* Panzer Regiment

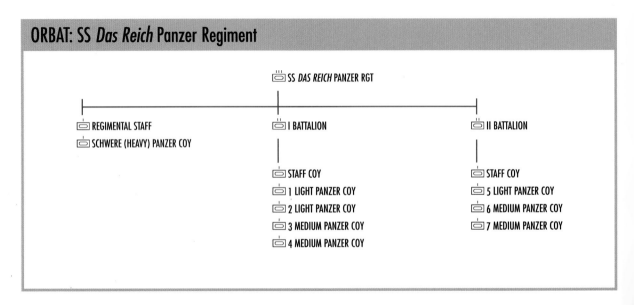

Panzer Regiment: Equipment

UNIT	EQUIPMENT	STRENGTH
Regimental HQ	Kübelwagen,	3
	Kettenkrads	4
HQ Company:	Kettenkrads,	2
	Kübelwagen,	1
	Panzerbefehlswagen III or IV (Panzer III or IV command tanks)	3
Reconnaissance Platoon:	Panzer III (HQ),	1
	Panzer III	4
Flak Platoon: HQ:	Motorcycle	2
	Kübelwagen,	1
3 x Flak sections, each with	Kettenkrad	1
	SdKfz 10/5	4
Tiger Company		
HQ:	motorcycle combination	1
	Motorcycle	2
	Kfz 15 radio car	1
	Panzer Mk VI Ausf E Tiger	2
3 x Platoons, each with	Panzer Mk VI Ausf E Tiger	3
Panzer Battalion 1		
HQ	Motorcycle	2
	Krupp Kfz 81 light trucks	3
	Kübelwagen	3
	Opel Blitz Type A	1
Battalion Company HQ:	Motorcycle	2
	Krupp Kfz 81 light truck	1
Panzer Platoon:	Panzer IIIM (HQ)	1
	Panzer IIIM	4
Signal Platoon:	Panzerbefehlswagen III or IV	1
	Opel Blitz 'Koffer' radio vehicle (HQ)	1
	Panzerbefehlswagen III or IV	2
Engineer Platoon:	motorcycle	1
	Kübelwagen	1
	Opel Blitz Type A (HQ)	1
4 x Scout Sections, each with	Motorcycle	2
	Kettenkrad	1
3 x Pioneer Sections, each with	SdKfz 251/7	1
	Maultier	1
AA Platoon:	motorcycle	1
	Kübelwagen (HQ)	1
3 x Sections, each with	SdKfz 7/1 with 20mm Flakvierling	1
	Maultier	1
2 x Panzer Companies, each with		
Company HQ:	Panzer Mk IIIM	2
	Motorcycle	2
	Krupp Kfz 81 light truck	1
3 x Platoons, each with	Panzer Mk IIIM	5

UNIT	EQUIPMENT	STRENGTH
2 x Companies, each with		
Company HQ:	Panzer Mk IVH	2
	Motorcycle	2
	Krupp Kfz 81 light truck	1
3 x Platoons, each with	Panzer Mk IVH	5
Panzer Battalion 2		
HQ	Motorcycle	2
	Krupp Kfz 81 light truck	3
	Kübelwagen	3
	Opel Blitz Type A	1
Battalion Company HQ:	Motorcycle	2
	Krupp Kfz 81 light truck	1
Panzer Platoon:	Panzer Mk IIIM (HQ)	1
	Panzer Mk IIIM	4
Signal Platoon:	Panzerbefehlswagen III or IV	1
	Opel Blitz 'Koffer' radio vehicle (HQ)	1
	Panzerbefehlswagen III or IV	2
Engineer Platoon:	motorcycle	1
	Kübelwagen	1
	Opel Blitz Type A (HQ)	1
4 x Scout Sections, each with	Motorcycle	2
	Kettenkrad	1
3 x Pioneer Sections, each with	SdKfz 251/7	1
	Maultier	1
AA Platoon:	motorcycle	1
	Kübelwagen (HQ)	1
3 x Sections, each with	SdKfz 7/1 with 20mm Flakvierling	1
	Maultier	1
1 x Panzer Company, with		
Company HQ:	Panzer Mk IIIM	2
	Motorcycle	2
	Krupp Kfz 81 light truck	1
3 x Platoons, each with	Panzer Mk IIIM	5
2 x Panzer Companies, each with		
Company HQ:	T-34 (747r)	2
	Motorcycle	2
	Krupp Kfz 81 light truck	1
3 x Platoons, each with	T-34 (747r)	3

Regimental HQ & HQ Company

Kübelwagen: 5
Kettenkrad: 9
Panzerbefehlswagen III: 8

Motorcycle: 2
SdKfz 10/5: 12

The regimental HQ and HQ Company consisted of mainly communications vehicles, in the shape of the Panzerbefehlswagen III (a version of the Panzer III armed with a short-barreled 50mm/1.97in gun), as well as a Flak platoon of 12 SdKfz 10/5, armed with 20mm (0.79in) automatic cannon.

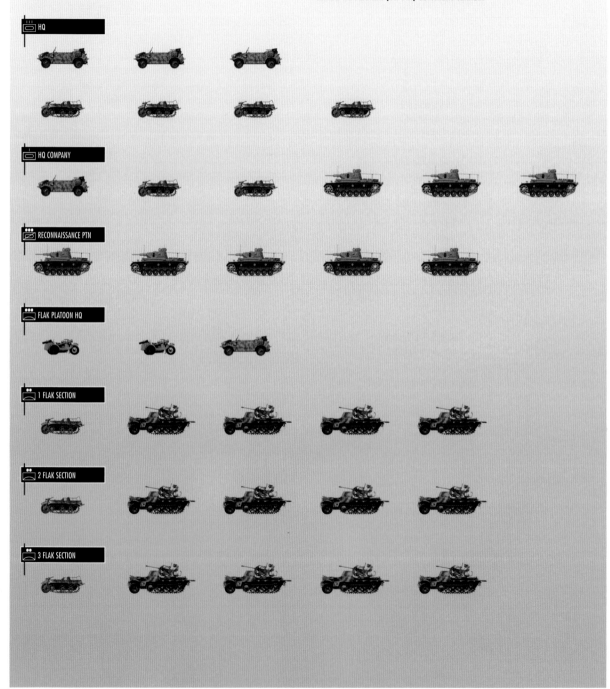

you from being surprised. Protect your flanks as well as your front. All-round observation is the duty of every commander. ALWAYS KEEP AN EYE OUT FOR THE ENEMY!

4. You must constantly evaluate the changing situation in combat. This is the only way in which you can make the correct decision in an emergency and issue short, clear orders without delay.

5. Strict radio discipline is essential for good leadership.

6. You must lead with strength. ... The more firepower which can be brought to bear in the first minute, the quicker the enemy will be defeated ...

7. When breaking cover, do it quickly and in unison. The more targets the enemy is faced with simultaneously, the harder his fire control will be, and the more firepower you can bring against him.

8. In the attack move as fast as possible. At low speed you are much more likely to be hit. In combat there should be only two speeds: slow for firing and flat out. This is a basic principal of tank combat!

9. When anti-tank weapons are encountered at long or medium ranges, you must first return fire and then manoeuvre against them. First halt to return fire effectively – then commit the bulk of the company to manoeuvre against the enemy, leaving one platoon to give supporting fire.

10. When anti-tank weapons are encountered at close range, stopping is suicidal. Only an immediate attack at full speed with all guns firing can be successful and reduce losses.

11. In combat against anti-tank guns you should never – even with the protection of strong fire support – allow a single platoon to attack alone. Anti-tank weapons are not employed singly. Remember – lone tanks in Russia are lost!

12. You must always continually keep your vehicles well spread out. This splits the enemy's defensive fire and complicates his fire control. Bunching must be avoided at all costs, especially in critical situations, or it will cost you casualties.

13. When an impassable obstacle such as a minefield or anti-tank ditch is encountered you must immediately and without hesitation give the order to withdraw to the nearest cover. In such circumstances halting in full view of the enemy or trying to carry on the attack will only cost you casualties. Your decision on what to do next is best made in the safety of cover.

14. When your attack must pass potential enemy tank positions, for instance a wood, you should either pass by them so closely that you are inside their minimum range, or keep so far away that you are outside their maximum effective range.

15. Enemy tanks should not be attacked directly, because they can see you and assess your strength before you can kill them. Avoid them until you can move into favourable firing positions and surprise them from flank or rear. Repelled enemy tank assaults must be aggressively pursued.

16. Whenever possible, a strongpoint should be attacked from different directions simultaneously in order to split the enemy's defensive fire and deceive him about the true source and direction of the attack. This will make your breakthrough easier and it will be achieved with fewer losses.

17. Always prepare dug-in positions and

A group of *Das Reich* panzer crew wait to unload a Panzer III at a railhead near their assembly point. The division's official marking, the *Wolfsangel*, can be seen on the hull of the tank.

2/2 SS Panzer Regiment, February 1943

EQUIPMENT	STRENGTH
Mk II	10
Mk III (lg)	81
Mk IV (lg)	21
Mk VI	10
Cmd	9

2/2 SS Panzer Regiment, July 1943

EQUIPMENT	STRENGTH
Mk II	1
Mk III (lg)	62
Mk IV (lg)	33
Mk VI	14
T-34	25
Cmd	10

camouflage against the possibility of air or artillery attack. Being sorry afterwards is no excuse for losses taken from these causes.

18. At decisive moments, do not try to conserve ammunition. At such times it is acceptable to expend ammunition at exceptionally high rates to minimize losses.

19. Never split your combat power – do not deploy elements of the company in such a way that they cannot support each other. When your attack has two objectives you should attack first one and then the other with your full strength. In this way you will more certainly take both objectives with fewer casualties.

20. Support from artillery fire or dive bombers must be used immediately, whilst the objective is still under bombardment. Do not wait until such attacks end, as they generally suppress rather than destroy the enemy. It is better to risk a friendly shell or bomb than to charge into an active anti-tank defence.

21. Do not misuse attached arms – for example, do not use tank destroyers as assault guns, or reconnaissance or engineers as infantry.

22. Protect non-armoured or lightly armoured units attached to you from any unnecessary losses until they are needed for the tasks for which they were attached.

23. Attached units are not your servants, but your guests and you are responsible for ensuring that they are properly supplied. Don't just use them for guard duty! In this way they will work better and more loyally for you when you need them. And that will be often!

24. In combined operations with infantry or armoured infantry, you must make certain that the arms work closely together to achieve success. Who takes the lead is a secondary matter; remember that the enemy will try to separate them and that you must prevent this at all costs. Your battle cry must be 'Protect the Infantry!' and the infantry's battle cry is 'Protect the Tanks!'

25. Always concentrate on your combat mission, and do not be diverted into attacking an enemy to your flank, unless he threatens the accomplishment of your mission.

26. After a victory always be prepared for the inevitable counter-attack.

27. In a defensive position, deploy your tanks so that they can make use of shock action as well their firepower. Only a few tanks should be in stationary firing positions, with the majority kept under cover to act as mobile reserves. Tanks defend aggressively!

28. There is no point in continuing to attack against strong enemy resistance. Every failed attack only costs more casualties. It is better to hold the enemy with weak forces, whilst you mass of your strength for a surprise attack from another quarter.

29. Never forget that your soldiers do not belong to you, but to Germany. Personal glory hunting and senseless dare-devilry lead only rarely succeed, but always cost blood. In battle against the Russians you must temper your courage with judgement, cunning and tactical ability. Only then will you earn the loyalty and respect of your soldiers.

30. The panzer division in modern warfare today holds the former place of cavalry as the decisive arm of combat. Tank officers must carry on the cavalry traditions and its aggressive spirit.

Remember Marshal Blucher's motto, 'FORWARD AND THROUGH!' (but sensibly).'

Panzer Battalion I

The Panzer IIIs that equipped *Das Reich* in the summer of 1943 were very different from the early versions which first went into action in Poland in September 1939.

The type originated in January 1934, when a requirement was issued for a medium tank with a maximum weight of 24 tonnes and a top speed of 35km/h (21.75mph). It was intended to be the battle tank of the panzer divisions, entrusted with the primary role of engaging and destroying opposing armoured forces.

PANZER III DEVELOPMENT
Daimler-Benz, Krupp, MAN and Rheinmetall all produced prototypes, and these were tested throughout 1936 and 1937. The Daimler-Benz design was eventually chosen for production. Even so, only 75 examples of the early models (designated Ausf A – Ausf D inclusive) were completed in 1937/38, all of which were essentially prototypes. Mass production of the type only really began in 1939, with the Ausf E and Ausf F versions.

All early versions up to and including the Ausf F were armed with the 37mm (1.45in) KwK 36, which proved to be virtually useless against well-protected Allied heavy tanks such as the Matilda and Char B, which were encountered in 1940. Fortunately for the *Panzerwaffe*, the Panzer III's turret ring was large enough to be upgunned with more potent 50mm (1.97in) weapons.

The first of these was the KwK 38 L/42 which, although offering a marked improvement in armour-piercing performance, was found to be woefully inadequate to deal with the T-34s and KV-1s in the opening stages of Operation *Barbarossa*. The gun that allowed the Panzer III to remain a viable battle tank well into 1943 was the 50mm (1.97in) KwK 39 L/60 and its tungsten-cored APCR ammunition. This combination could penetrate the T-34's frontal armour at up to 500m (1640ft) and defeat the KV-1's side armour at similar ranges. (Although this solved

PzKpfw III Ausf M

■ CREW

■ SPECIFICATIONS

Weight: 21.13 tonnes (20.8 tons)
Length: 6.412m (21ft)
Width: 3.266m (10ft 8.5in)
Height: 2.5m (8ft 2.4in)
Engine: 223.7kW (300hp) Maybach HL120TRM V12 petrol
Speed: 40km/h (25mph)
Range: 155km (95 miles)
Armament: 1 x 50mm (1.97in) KwK39 L/60, plus 2 x 7.92mm (0.31in) MG34 machine guns

The Ausf M was one of the last variants of the Panzer III to be produced as a conventional tank. Its armour had been improved to a point where it was very nearly as good as contemporary Panzer IVs, and its main armament was still good enough to deal with the majority of Soviet tanks encountered at Kursk. However, it was withdrawn from frontline service in late 1943, as it was too small to be up-gunned to meet the challenge of the Red Army's newer AFVs.

HQ: Panzer Battalion 1

Motorcycle: 14
Kfz 81 light truck: 4
Kübelwagen: 5

Panzerbefhlswagen III: 8
SdKfz 7/1 flak: 3
Opel Bltiz truck: 3

SdKfz 251/7: 3
Maultier: 6
Kettenkrad: 4

HQ

COMPANY HQ

1 ZUG

SIGNAL PLATOON

ENGINEER PTN HQ

1 SCOUT SECTION   **3 SCOUT SECTION**

2 SCOUT SECTION   **4 SCOUT SECTION**

1 PIONEER SECTION   **2 PIONEER SECTION**   **3 PIONEER SECTION**

AA PLATOON HQ

1 AA SECTION   **1 AA SECTION**   **1 AA SECTION**

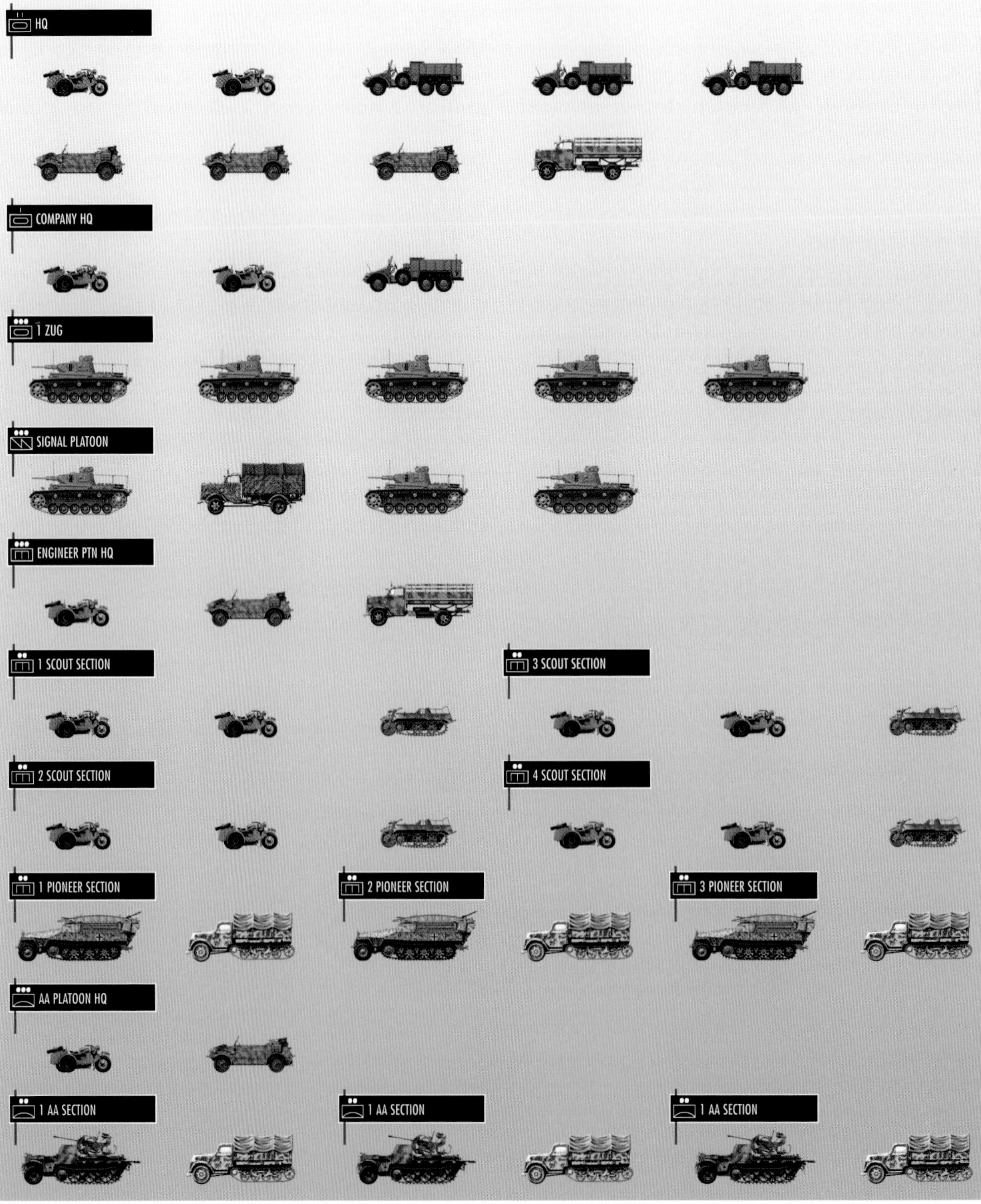

one problem, it caused another, as German stocks of tungsten were already low and in great demand for machine tools. The situation became so critical in late 1943 that production of APCR ammunition was halted to conserve remaining tungsten supplies for German war industries.)

The final gun to be successfully fitted to the Panzer III was the 75mm (2.95in) KwK 37 L/24, which was available in quantity – surplus stocks had accumulated as Panzer IVs were rearmed. Attempts were made to extend the type's usefulness by fitting the 75mm (2.95in) KwK 40 L/43, but it would have required a major redesign of the vehicle to enlarge the turret ring sufficiently to take such a powerful weapon and it was felt that future Panzer III hulls were better utilized as the basis for assault guns.

The rearmed Panzer IIIs were also fitted with thicker armour – the earliest models had a maximum of only 15mm (0.5in) which was steadily improved in successive models to reach 70mm (2.75in) in the Ausf J, L, M and N, which were the main variants still in service in mid-1943. This was sufficient to give virtual immunity to the Red Army's 45mm (1.8in) tank and anti-tank guns except at point-blank ranges, plus a high degree of protection from the 76.2mm (3in) guns of the T-34 and KV-1 at ranges above 500m (1640ft).

In June 1942 there were still approximately 1100 Panzer IIIs in first line service in Russia and the type played an important role in the last significant German victories on the Eastern Front in early 1943. It was still an effective weapon at this time, but by the summer of 1943 it was becoming increasingly outclassed. In July 1943 Army Groups Centre and South fielded a total of 432 Panzer IIIs with the L/60 gun for Operation 'Citadel', but this was the last time that the type was deployed in such large numbers. Production of the Panzer III finally ended in August 1943, and by early 1944 most surviving vehicles had been relegated to training or counter-insurgency duties, in which they continued to serve until the end of the war.

PANZER IV

The Panzer IV was developed in parallel with the Panzer III. In January 1934 a requirement was issued for a tank armed with a 75mm (2.95in) howitzer to supplement the Panzer III. The type was

to be optimized for the close support role, targeting enemy anti-tank guns and field fortifications to allow the Panzer IIIs to deal with opposing AFVs. MAN, Krupp and Rheinmetall-Borsig each developed prototypes, with Krupp's being selected for production. Minor improvements were incorporated into successive models, but production was relatively slow and less than 500 vehicles of all versions had been completed by the summer of 1940.

All these were armed with the 75mm (2.95in) KwK 37 L/24, a short barrelled gun primarily designed for firing HE, but with a secondary anti-tank capability. Even with the excellent sights fitted to all Panzer IVs, it took considerable skill to hit a moving target with such low velocity weapon, but the superbly trained crews of 1939/40 managed well enough. Initially, the armour-piercing round was a conventional APCBC shell capable of penetrating 46mm (1.8in) of armour at up to 500m (1640ft),

Panzer Battalion 1: Equipment		
UNIT	EQUIPMENT	STRENGTH
1 Company HQ	Panzer IIIM	2
	Motorcycle	2
	Krupp Kfz 81 light truck	1
1 Platoon	Panzer IIIM	5
2 Platoon	Panzer IIIM	5
3 Platoon	Panzer IIIM	5
2 Company HQ:	Panzer IIIM	2
	Motorcycle	2
	Krupp Kfz 81 light truck	1
4 Platoon	Panzer IIIM	5
5 Platoon	Panzer IIIM	5
6 Platoon	Panzer IIIM	5
3 Company HQ	Panzer IVH	2
	Motorcycle	2
	Krupp Kfz 81 light truck	1
7 Platoon	Panzer IVH	5
8 Platoon	Panzer IVH	5
9 Platoon	Panzer IVH	5
4 Company HQ	Panzer IVH	2
	Motorcycle	2
	Krupp Kfz 81 light truck	1
10 Platoon	Panzer IVH	5
11 Platoon	Panzer IVH	5
12 Platoon	Panzer IVH	5

Panzer Battalion 1

Motorcycle: 8
Kfz 81 light truck: 4
PzKpfw III Ausf M: 34

PzKpfw IV Ausf H: 34

The SS Panzer battalions were incredibly powerful formations, consisting of up to 70 Panzer III and IV tanks. Each battalion included four companies, with three platoons, or *Zug* – a *Zug* comprised five panzers. A battalion would be easily capable of taking on and beating a Soviet tank brigade of the period.

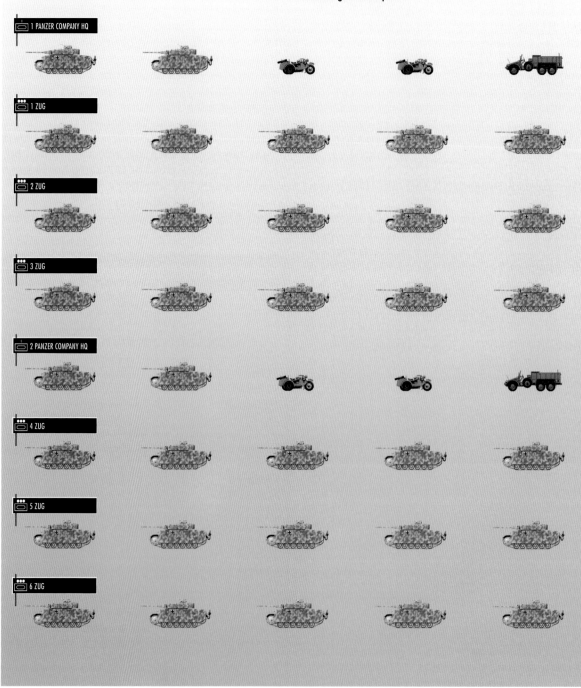

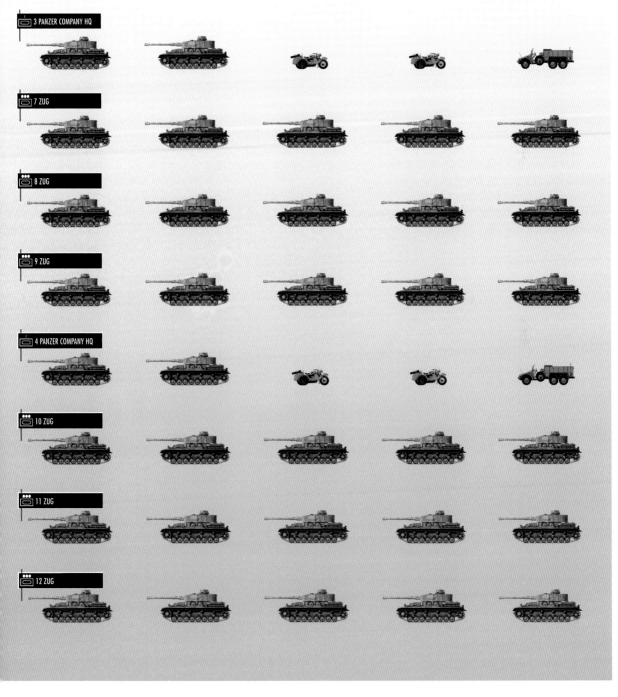

3 PANZER COMPANY HQ

7 ZUG

8 ZUG

9 ZUG

4 PANZER COMPANY HQ

10 ZUG

11 ZUG

12 ZUG

which was thought to be adequate to deal with all known Allied tanks. This illusion was shattered by actions in France during the summer of 1940, when the near- invulnerability of the Matilda and Char B sparked off a series of frantic efforts to improve the Panzer IV's anti-tank capability.

NEW ROUNDS

The first measure was the introduction of a series of HEAT rounds for the KwK37 – the earliest types had a maximum penetration of 70mm (2.75in), but later versions could penetrate 100mm (3.9in). Whilst this helped, it was recognized that a higher velocity weapon was really required for the anti-tank role and a 50mm L/60 gun based on the 50mm PaK 38 anti-tank gun was experimentally fitted to a Panzer IV Ausf D. However, with the rapid German victory in France, the initial order for 80 such tanks was cancelled before production began.

Little serious work was carried out on up-gunning the design until November 1941, when the flood of reports from the Eastern Front complaining about the Panzer IV's ineffectiveness against the T-34 and KV-1 compelled urgent action. Any thought of up-gunning the type with

the 50mm L/60 gun was dropped and Krupp was brought in to modify Rheinmetall's new 75mm (2.95in) anti-tank gun (later known as the PaK 40 L/46). As the recoil length of the PaK 40 was too great for the tank's turret, the recoil mechanism and chamber were adapted to produce the 75mm KwK40 L/43.

The new gun was first fitted to the Panzer IV Ausf F2, which entered service in March 1942. Coupled with improved armour protection, it brought the type back to near-parity with the T-34 and gave it the ability to destroy even the KV-1 with flank or rear shots at ranges up to 500m (1640ft). In June 1942, the longer barrelled, higher velocity 75mm KwK40 L/48 was introduced in the new Ausf G, which, together with further up-armouring, made this version at least equal to the T-34.

A total of 841 Panzer IVs were deployed for the Kursk offensive and were arguably the most important AFVs in the German forces at that time, with far better firepower than the Panzer III and much greater battlefield mobility and reliability than the Tiger or Panther. Total production of all versions exceeded 8500 and the Panzer IV remained in service throughout the war.

PzKpfw IV Ausf H

■ CREW

■ SPECIFICATIONS

Weight: 26 tonnes (25.6 tons)
Length: 7.015m (23ft)
Width: 3.33m (10ft 11in)
Height: 2.68m (8ft 9.5in)
Engine: 223.7kW (300hp) Maybach HL120TRM V12 petrol
Speed: 38km/h (24mph)
Range: 210km (130 miles)
Armament: 1 x 75mm (2.95in) KwK40 L/48, plus 2 x 7.92mm (0.31in) MG34 machine guns

The latest version of the Panzer IV to equip *Das Reich* at Kursk was the Panzer IV Ausf H, which entered production in April 1943. Its frontal armour was increased to a maximum of 80mm (3.15in) and Zimmerit paste was applied to prevent the adhesion of magnetic anti-tank mines, which the Germans feared would be used in large numbers by Allied infantry. The vehicle's sides and turret were further protected by the addition of *Schurzen* – 5mm (0.2in) side-skirts and 8mm (0.31in) turret skirts.

Heavy Panzer (Tiger) Company

Between 1937 and mid-1941, Henschel and Porsche had been steadily developing a series of medium and heavy tank designs. Although the heavy British and French tanks encountered in 1940 had proved to be almost invulnerable to the fire of Panzer IIIs and IVs, there was little sense of urgency in getting new types into production.

On 26 May 1941, Hitler ordered Porsche and Henschel to produce prototypes of a new 45-ton tank armed with the 88mm (3.5in) L/56. The threat posed by Soviet T-34s and KV-1s became apparent in the first weeks of Operation *Barbarossa* and gave the project a far greater sense of urgency. Porsche had an initial advantage in the race to win the production contract because his design studies were much more advanced than Henschel's, but he chose to adopt a petrol-electric drive system. This comprised two air-cooled 238.62kW (320hp) Type 101/1 petrol engines driving generators, which in turn powered the two electric motors that actually drove the tracks. Although it was a theoretically fuel-efficient system, it proved to be complex and unreliable, besides using large quantities of scarce copper. Henschel wisely utilized a much more conventional Maybach petrol engine, which proved to be far less prone to breakdowns.

Both prototypes were fitted with almost identical Krupp turrets armed with the 88mm (3.5in) KwK36 L/56 and a co-axial MG before being inspected by Hitler in April 1942. They then underwent competitive trials, which ruthlessly exposed the Porsche design's inherent unreliability and poor cross-country performance. Accordingly, the Henschel design was ordered into production in August 1942, officially becoming the Panzerkampfwagen Tiger Ausf E, although to most contemporaries, it was simply the Tiger.

THE TIGER I

When it began mass production of the Tiger I in August 1942, the Henschel factory at Kassel already had eight years experience of manufacturing AFVs, although the massive Tiger was very different to the little Panzer Is which it had started building in 1934.

During the war years, the factory employed a total of 8000 workers in AFV production and

Tiger Company: Equipment

UNIT	EQUIPMENT	STRENGTH
Company HQ:	Motorcycle combination	1
	Motorcycle	2
	Kfz 15 radio car	1
	Panzer Mk VI Ausf E Tiger	2
1 Platoon	Panzer Mk VI Ausf E Tiger	3
2 Platoon	Panzer Mk VI Ausf E Tiger	3
3 Platoon	Panzer Mk VI Ausf E Tiger	3

operated around the clock, working two 12-hour shifts. Each six-hour stage in the manufacturing process was referred to as a *takt* and there were nine such *takte* in producing the Tiger I. The total time to complete a Tiger, including the various machining processes, was estimated to be 14 days. At any one time, an average of 18–22 tanks were on the hull assembly line with a further 10 tanks on the final assembly line. (It has been estimated that the production of each Tiger took 300,000 man-hours.)

Henschel did not have the capability to weld or bend the massive armour plates used in the Tiger, and received the hull and turret shells from other companies. The hulls were manufactured by two firms, Dortmund-Hoerder Huettenverein and Krupp, whilst the turrets were produced by Wegmann und Company, which was also based in Kassel.

In common with all other panzers, the Tiger I was continually modified throughout its production run. Problems caused by mud, ice and snow jamming the interleaved road wheels were first encountered by Tiger units in Russia during the winter of 1942/43, and this was not solved until the introduction of the Tiger II which had overlapping, rather than interleaved, road wheels. However, the process of improving the type began almost immediately after production had started in August 1942 – the 'pre-Kursk' updates included:

August 1942
- Mounting for camouflage frame added to hull
- Smoke bomb dischargers fitted to the turret sides

November 1942
- Feifel air cleaner system installed
- Track guards fitted to the hull sides with hinged front and rear plates
- Reinforced gun mantlet fitted

December 1942
- An emergency escape hatch added to the turret

January 1943
- A total of five S-Mine dischargers installed – one forward-firing and one to each corner of the hull. These had a range of only a few metres, but were deadly against attacking infantry because the detonation of each mine scattered a total of 360 steel balls lethal at up to 20m (65ft 7in) and capable of wounding anyone within 100m (328ft).
- A purpose-designed turret stowage bin fitted

March 1943
- Modifed Feifel air filters introduced

- Loader's periscope fitted to turret roof

May 1943
- 521.99kW (700hp) Maybach HK 230 P45 engine with two air filters replaced the 484.7kW (650hp) Maybach HL 210 P45
- Improved transmission installed

IN COMBAT

The first Tiger Is were issued to the 1st Company, 502nd *schwere Panzer Abteilung* (Heavy Panzer Battalion) in August 1942. The unit was attached to Army Group North for operations on the Leningrad sector of the front, where the terrain posed almost as many problems as the enemy.

General Heinz Guderian wrote scathingly of this deployment in his book *Panzer Leader:*

'He (Hitler) was consumed by his desire to try his new weapon. He therefore ordered that the Tigers be committed in a quite secondary operation, in a limited attack carried out in terrain that was utterly unsuitable; for in the swampy forest near Leningrad heavy tanks could only move in single file along the forest tracks, which, of course, was exactly where the enemy anti-tank guns were

Tiger Company: Operational Strength, 5 July 1943

Motorcycle: 3
Kfz 15 radio car: 1
PzKpfw VI Ausf E Tiger I: 11

Even during quiet periods, it was extremely rare for any panzer company to have all its tanks fully operational. The Tiger's weight and sophistication made it particularly prone to mechanical problems and on 5 July only 11 of *Das Reich's* 14 Tigers were operational.

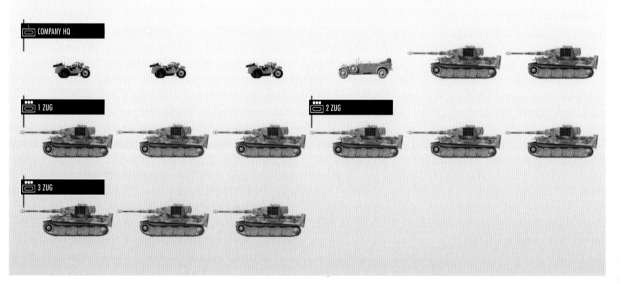

posted, waiting for them. The results were not only heavy, unnecessary casualties, but also the loss of secrecy and of the element of surprise for future operations.'

During this first attack in late August all the Tigers sustained some damage and one was captured. Further attacks over the next month or so were equally unsuccessful. The Tiger's undoubted technical superiority to any contemporary AFV in the Red Army's inventory was wasted in offensive operations in this setting, where all the advantages lay with well-camouflaged anti-tank guns deployed in depth to cover the few roads in the area. (Although the Tigers' armour was virtually invulnerable to Soviet anti-tank guns, most of their losses were to 'mobility kills' – vehicles were immobilized by shots breaking tracks or damaging the running gear.)

When the Soviets went over to the offensive in this sector early in 1943, the Tigers were in turn able to exploit the swampy forest terrain that forced most Soviet AFVs to keep to the roads. During a period of less than three months (12 January to 31 March 1943) the company lost 6 Tigers, whilst destroying 160 Soviet tanks – a ratio of 26:1.

In common with most other heavy panzer battalions, the unit was plagued by inadequate recovery equipment and a constant shortage of spares, which kept many of their Tigers unserviceable. The unit never had more than four operational Tigers at any one time during the entire period. Three of the six Tigers lost were destroyed by their own crews to prevent capture; two of them after they had become stuck in swamps; and one due to mechanical failure. (The unit's war diary is filled with entries about retrieving 'bogged-down' Tigers – in one case, the recovery operation took three days to complete.)

WAFFEN-SS TIGERS

Himmler's influence ensured that the *Waffen-SS* had a high priority for the issue of the new Tigers. On 15 November 1942, three *schwere Panzer Kompanien* (Heavy Panzer Companies) were established, one each for *Leibstandarte, Das Reich* and *Totenkopf*. Each company was to have nine Tigers and ten Panzer III.

PzKpfw VI Ausf E Tiger I

The popular image of Prokhorovka is one of massed Tigers fighting hordes of T-34s, but in reality, the three divisions of II SS Panzer Corps had a total of just 15 operational Tigers on the morning of 12 July, only one of which belonged to *Das Reich*. Before the Kursk offensive, the Tiger company changed its vehicle numbering system. The first digit on each tank was changed from an '8' to an 'S' for *schwere* (heavy), followed by a two digit tank number denoting platoon and vehicle.

■ CREW

■ SPECIFICATIONS

Weight: 57 tonnes (56.1 tons)
Length: 8.432m (27ft 8in)
Width: 3.7m (12ft 1.6in)
Height: 3m (9ft 11in)
Engine: 522kW (700hp) Maybach HL230P45 V12 petrol
Speed: 45km/h (28mph)
Range: 195km (120 miles)
Armament: 1 x 88mm (3.5in) KwK36 L/56, plus 2 x 7.92mm (0.31in) MG34 machine guns

A total of 28 Tigers and 30 Panzer IIIs were issued in December 1942 and January 1943. All three companies were sent to the Eastern Front, taking part in von Manstein's counter-offensive to retake Kharkov in February/March 1943, when they lost a total of five Tigers.

An order dated 22 April 1943 authorized these companies to be changed to 14 Tigers each. The same order officially merged the companies into an *SS schwere Panzer Abteilung* (SS Heavy Panzer Battalion). In practice, however, the three companies remained with their Regiments at the front and in May 1943, they received a further 17 Tigers, bringing the unit totals to:

- *Leibstandarte* – 13 with 13.sKp/SS-PzRgt.1
- *Das Reich* – 14 with sKp/SS-PzRgt.2
- *Totenkopf* – 15 with 9.Kp/SS-PzRgt.3

Operations in February/March 1943 were highly successful, in part because the Tigers' battlefield mobility was improved by the bitter cold which froze streams and marshes. (At other times of year, these were always the most difficult obstacles for Tiger units – lighter panzers could cope with them more readily, but the Tiger I's combat weight of 57 tonnes/56.1 tons always made it liable to bogging down in such terrain.)

Much also depended on the efficiency of divisional recovery units – immobilized Tigers required careful handling and attempts to use another Tiger as a recovery vehicle frequently led to two broken down vehicles instead of one. The recovery teams' 18 tonne (17.7-ton) SdKfz 9 halftracks had a nominal towing capacity of 28 tonnes (27.55 tons) and, depending on the terrain, two or even three might be required to haul a single Tiger back to the repair centre.

The meticulous *Obergruppenführer* Paul Hausser commanding the SS Panzer Corps ensured that such lessons learnt by the first Tiger units were applied throughout the Corps. Experience showed that Tiger platoons could effectively open fire (concentrated platoon fire) against stationary targets at up to 3000m (9843ft). When firing against moving targets, it became general practice to open fire at between 1200m (3937ft) and 2000m (6562ft) depending on the tactical situation. These tactics were applied effectively by the *Waffen-SS* Tiger companies throughout Operation 'Citadel', but other units were not so efficient.

General Breith, commander of the III Panzer Corps, was certainly unhappy with he way in which his Tigers had been deployed and issued the following directive on 21 July 1943:

'Based on experience in the recent battles, I issue the following instructions for the cooperation of Tigers with other weapons:

1. As a result of its high performance weapon and strong armour, the Tiger should be used primarily against enemy tanks and anti-tank

Chronology: *Das Reich's* Tiger Company

DATE	EVENT
15 November 1942	Formation of unit ordered as the Tiger company of 2 SS-Panzergrenadier Division *Das Reich*
December 1942	2 Tigers received
January 1943	8 Tigers and 12 Panzer IIIN received. These are organized as four heavy platoons, each with 3 Tigers and 1 Panzer IIIN and two light platoons, each with 4 Panzer IIIN
24/25 January 1943	Company entrained for the Eastern Front
1/2 February 1943	Arrival in Kharkov
10 February 1943	Tiger commanded by *Obersturmführer* Gerlach is captured after being immobilized by a mine
17 February 1943	Withdrawal to Poltava begins – replacement for captured Tiger is received
22 February 1943	Company participates in von Manstein's counter-offensive
25 February 1943	3 Tigers operational
1 March 1943	An immobilized Tiger is destroyed to prevent its capture
13 May 1943	6 Tigers delivered, 14 now operational
20 May 1943	Remaining Panzer IIIN transferred
4 July 1943	12 Tigers operational
5 July 1943	Operation 'Citadel' begins – 11 operational Tigers
7 July 1943	7 Tigers operational
8 July 1943	6 Tigers operational
9–11 July 1943	1 operational Tiger
12 July 1943	2 Tigers operational
13 July 1943	1 operational Tiger
14 July 1943	4 Tigers operational
15 July 1943	2 Tigers operational
16 July 1943	5 Tigers operational
17–18 July 1943	9 Tigers operational

weapons and secondarily – and then only as a complete exception – against infantry targets. As experience has shown, its weapons allow the Tiger to fight enemy tanks at ranges of 2000 metres and longer, which has especially worked on the morale of the opponent. As a result of the strong armour, it is possible to close to short range with the enemy tanks without being seriously damaged from hits. Still, the Tiger should not attempt to start enemy tanks at ranges over 1000 metres.

As often as the situation allows – which was possible very often during the recent battles – prior to employment of Tigers the terrain should be scouted with the primary purpose of determining the possibility of crossing rivers and streams, bridges and marshlands. Kompanie and Zugleaders and also Panzer commanders may not shy from dismounting and performing scouting patrols on foot in order to prevent the entire unit from getting stuck in difficulty terrain. In connection with this, unnecessary losses of Tigers on mines could have been prevented. The same applies to the other types of Panzers.

A known weakness of the Tiger is caused by the location of the commander's cupola on the left side of the turret. The commander can't see an extensive area close to the right side of the Tiger, which presents a threat from the opponent's tank hunter teams. Therefore, it is well known that it is necessary for other troops to protect the Tiger from this threat.

2. During the attack on 5 July in a sector of the Korps front, lack of knowledge about our own minefields was detrimental to our Tigers. Therefore, the forward Kampfgruppen should plan to have sufficient Pioniere not only to clear away obstacles but also to clear minefields.

3. I forbid employment of Panzers, especially Tigers, in less than Kompanie strength. For defence, Panzers are to be consolidated into attack groups that are to be sent in as planned counter strikes. After completing a counter strike, these Panzer groups are to be immediately pulled back to return to the disposal of the sector or division commander. Dispersal of Panzers in the main battle line or guarding other weapons by day and by night must not occur.'

3/2nd SS Panzer Regiment, July 1943

The departure of the Panzer Regiment's 1st Battalion for re-equipping with the Panther left the regiment seriously under-strength. As a temporary measure, a 3rd Battalion was formed, which was equipped with a mixture of Panzer IIIs and captured T-34s.

German forces had made extensive use of captured AFVs since early 1939, when the Czech LT vz. 35 and TNHPS tanks were taken into service as the Panzer 35(t) and Panzer 38(t). These types were vitally important in 1939–41, when the 1600+ vehicles in service formed a high percentage of the *Panzerwaffe's* strength. Throughout the war years, army and *Waffen-SS* units were equipped with a wide range of foreign AFVs that had been captured or produced in occupied territories under German supervision.

Captured tanks, designated *Beute Panzerkampfwagen* (Booty Panzers) were gathered at special collection points for detailed technical examination to determine if they were of any use. The best would be assigned for combat use, but even totally obsolete vehicles and those in very poor condition could still be used a targets on gunnery ranges.

The large numbers of captured T-34s were something of a special case – those which went into action in the opening stages of Operation *Barbarossa* were terrifying to German units, which were shaken by their firepower, cross-country mobility and their apparent invulnerability to the majority of German AFV guns and anti-tank weapons. This was especially apparent in the case of the German standard anti-tank gun of 1941, the 37mm (1.45in) PaK 36. Combat reports included that from a particularly determined (and lucky!) crew of a PaK 36, who fired 23 rounds at a single T-34 and succeeded only in jamming the turret. Whilst this highlights the effectiveness of the T-34's armour on the 1941 battlefield, it also emphasizes the tank's abysmal fire control system, which allowed the gunners to fire 23 rounds and survive without taking a single hit or damaging near miss in return.

Several factors contributed to this situation, including:

■ The two-man turret of the T-34, in which the grossly over-worked tank commander also acted as the gunner. If he was also a platoon or company commander, he had the additional daunting task of controlling several other tanks. The command and control problems were worsened by the dire shortage of radios, which affected Soviet armoured units for much of the war.

■ The poor quality of the tank's sighting and observation equipment. There was no commander's cupola and he had to rely on a single periscopic sight for observation when the tank was operating 'closed down'. (Soviet periscopic and telescopic sights were markedly inferior to their German counterparts, whilst the armoured glass blocks used in vision ports were often cloudy and full of bubbles.)

■ Any attempt by the commander to overcome this problem by operating with the turret open was hindered by the design of the hatch. This was a single large forward-opening hatch covering almost the entire rear half of the turret. Its weight and awkwardness were hated by Soviet tank crews who dubbed it *pirozhok* (stuffed bun) after its characteristic shape. When opened, it stood almost upright and blocked the commander's forward vision, forcing him to lean out to one side to get a clear view. If it jammed, the turret crew were trapped, with only a slim chance of being able to escape through the driver's hatch. Tank commander Nikolai Evdokimovich Glukhov remembered that it was: 'A big hatch – very inconvenient, very heavy.' The flood of complaints from tank crews finally led to its replacement with twin hatches in the Model 1943 turret.

■ Up to and including the Model 1942, the turret was very cramped and so low that the maximum depression of the 76.2mm (3in) and co-axial machine gun was restricted to only 3°. This severely limited the tank's ability to take up hull-down firing positions on reverse slopes and to fire on close-range targets.

Captured T-34s in good condition were modified to eliminate as many of these problems as possible. These modifications included:

■ The replacement of the single large turret hatch with a new roof section fitted with a commander's cupola taken from a damaged Panzer III or Panzer IV and a loader's hatch. This dramatically improved the commander's field of vision and gave the turret crew a reasonable chance of escape if the tank was knocked out.

■ Installation of German radio equipment, usually the FuG5.

■ Fitting additional stowage to give the crew slightly greater room – this varied from tank to tank.

Das Reich: T-34 establishment

DATE	SERVICEABLE
4 July	18
5 July	16
7 July	15
8 July	14
9 July	7
10 July	7
11 July	8
12 July	8
13 July	11
14 July	12
15 July	13
16 July	11
17 July	17
18 July	17

LEFT: Although Das Reich had a total of 25 captured T-34s immediately before Operation 'Citadel' began, no more than 18 were operational at any one time. This compared well with the serviceability rates for standard German AFVs, especially as some spares could only be obtained by cannibalising knocked-out Soviet T-34s. These were relatively easy to obtain whilst Das Reich was advancing or holding its positions against Soviet armoured attacks, but the supply dried up rapidly during the retreats following the Kursk offensive, when the divisional recovery teams rarely had the opportunity to strip immobilised Red Army tanks.

On 4 July 1943, *Das Reich's* order of battle included 18 T-34s, such as the one shown here. To avoid battlefield confusion, they were not used in the first wave of the assault and were clearly marked with German insignia.

Sometimes sheet metal boxes were fitted to the hull sides, whilst turret stowage bins salvaged from damaged Panzer IIIs and IVs were also frequently used.

◼ *Schurzen* (armoured skirting plates) were added to a number of T-34s to protect the suspension and top run of the tracks from anti-tank rifle fire.

◼ As with all captured AFVs, T-34s were repainted in German camouflage and oversize national markings to minimize the considerable risk from friendly fire. Large swastika flags were frequently draped over the rear decking to serve as air recognition markings.

The substantial numbers of captured T-34s led to the establishment of a workshop in Riga in late 1941, which specialized in their repair and modification. By 1943, Mercedes-Benz in Marienfelde and Wumag in Goerlitz (now Zgorzelec) were also repairing and modifying T-34s. Spare parts were never much of a problem as a total of 300 captured vehicles were usually held in storage at any one time. A number of captured T-34s were also used as artillery tractors and ammunition carriers.

Even quite badly damaged vehicles were used in action – some were dug in as pillboxes whilst others had their turrets removed and mounted on railway wagons designated as *Panzerjägerwagen* (tank destroyer wagons), which were incorporated into several armoured trains.

PzKpfw T-34 747(r)

In March 1943, II SS Panzer Corps recaptured Kharkov and seized a factory where at least 50 T-34s were under repair. The installation was pressed into service as an *SS Panzerwerk* (SS Tank Workshop) and the T-34s were modified to German standards. The modifications varied from tank to tank, but included installation of commander's cupolas (from damaged PzKpfw III and IV tanks), *Schurzen* (armoured skirts) and other equipment such as Notek lights, storage boxes, tools and radio equipment. A total of 25 were taken into service with *Das Reich's* Panzer Regiment.

◼ CREW

◼ SPECIFICATIONS

Weight: 30.9 tonnes (30.41 tons)
Length: 6.75m (22ft 1in)
Width: 3m (9ft 8in)
Height: 2.45m (8ft 0in)
Engine: 373kW (500hp) V-2 diesel
Speed: 55km/h (34.2mph)
Range: 465km (288.95 miles)
Armament: 1 x 76.2mm (3in) F-34 gun plus 2 x 7.62mm (0.3in) DT machine guns

2nd SS Flak Battalion

German forces were amongst the first to appreciate the potential impact of close air support on the battlefield and rapidly developed powerful AA units, which were both highly effective in their intended role and capable of rapid deployment against ground targets.

At full strength, the flak battalion could deploy 8 x 88mm (3.5in) guns and no less than 30 x 20mm (0.79in) Flakvierling quadruple cannon. This was a powerful concentration of highly mobile firepower (24 of the Flakvierling were self-propelled), which was devastatingly effective against even the well-armoured *Shturmoviks*. The '88s' were equally deadly when used in the anti-tank role, whilst the Flakvierlings were in great demand as infantry support weapons due to their accuracy and high rate of fire.

8.8CM FLAK 18, 36 AND 37

The first prototypes of the 88 were completed in 1928. These early models, the FlaK 18, had a single-piece barrel with a length of 56 calibres, hence the designation 88/L56. Series production began in 1933 and small numbers were sent to Spain for combat trials with the Condor Legion during the Spanish Civil War.

The FlaK 18 was mounted on a 360° traverse cruciform gun carriage. Two of its 'side' legs could be quickly folded up, allowing the gun to be lifted onto two single axle trailers for high-speed towing. Despite the gun's weight and bulk, which required a powerful prime mover such as the SdKfz 7 halftrack, a well-practised crew could remove the wheels and be ready to fire in 20 seconds and

remount the gun to move off in less than a minute. The semi-automatic breech allowed a rate of fire of 15–20rpm, which was better than most of its foreign counterparts of the era. The type proved its effectiveness in the Spanish Civil War, both as one of the best AA weapons of the time and as an outstanding long-range anti-tank gun.

The FlaK 18 was clearly capable of further development and a number of improvements were made to produce a successor, the FlaK 36. This had a two-piece barrel for easier replacement of worn liners, and new, heavier trailers that simplified the procedure for bringing it into action. In common with the FlaK 18, the type was often fitted with an armoured shield to give limited protection for the gun crew. A further version, the FlaK 37, included updated fire control equipment. The parts of all versions of the guns were interchangeable, and it was not uncommon for parts from different versions to be 'mixed and matched' on any given weapon.

■ **8.8 cm FlaK 18** A semi-automatic breech, high-velocity gun. Krupp began series production in 1933. Used the Sonderanhänger 201 trailer. Had a rate of fire of 15–20rpm. Later fitted with a shield to protect the crew when they were engaging ground targets.

■ **Mod 1938 II** Approximately 50 guns modified so

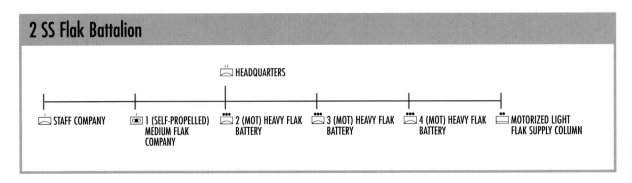

2 SS Flak Battalion

HEADQUARTERS

STAFF COMPANY — 1 (SELF-PROPELLED) MEDIUM FLAK COMPANY — 2 (MOT) HEAVY FLAK BATTERY — 3 (MOT) HEAVY FLAK BATTERY — 4 (MOT) HEAVY FLAK BATTERY — MOTORIZED LIGHT FLAK SUPPLY COLUMN

Flak Battalion HQ with Signals Platoon

Kübelwagen: 7   Kfz 17 radio car: 8
Motorcycle: 1   Opel Blitz trucks: 2
Kettenkrad: 1

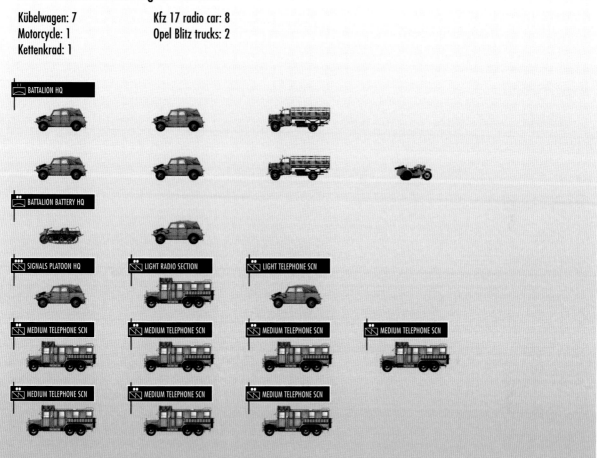

that a single gunner could control elevation and traverse.

■ **8.8 cm FlaK 36** Entered service 1936/37. It used the redesigned trailer Sonderanhänger 202, allowing faster time into action and the ability to engage ground targets whilst still in the travelling position. Weight 7 tonnes. Rate of fire 15–20rpm. Later fitted with a shield to protect the crew when engaging ground targets.

■ **8.8 cm FlaK 37** An updated version of the FlaK 36, the main difference being the introduction of the Übertragungser 37 (a new fire AA control system).

2CM FLAK 30/38/FLAKVIERLING

The FlaK 30 design originated with the Swiss Solothurn ST-5 20mm (0.79in) cannon, which was initially developed into the C/30 for the *Kriegsmarine*. The gun fired the 'Long Solothurn', a 20 x 138mm belted cartridge that had been developed for the ST-5 and was one of the most powerful 20mm (0.79in) rounds in existence.

In the early 1930s, Rheinmetall adapted the C/30 for Army use, producing the 2cm FlaK 30. This was similar to the C/30, with an entirely new mount – the gun was set up by sliding it to the ground from its two-wheeled carriage and levelling with hand cranks, creating a stable triangular base that allowed for a traverse of 360°.

The main problem with the gun was its rate of fire, which at 120rpm was not particularly high for a weapon of this calibre. Rheinmetall accordingly developed the 2cm FlaK 38, which entered service in 1939. This was a very similar weapon, but with

Heavy Battery

Kübelwagen: 9
Motorcycle: 11
Maultier: 3
Kfz 17/2 radio van: 5

SdKfz 7: 1
Opel Blitz truck towing 3 x 20mm Flakvierling: 3
SdKfz 7 towing 2 x 88mm Flak: 4

The flak battalion's 8 x 88mm (3.5in) guns and 3 x 20mm (1.79in) Flakvierling quadruple 20mm cannon offered a powerful concentration of highly mobile firepower, which was devastatingly effective against even the well-armoured *Shturmoviks*. The '88s' could also be used in the anti-tank role, whilst the Flakvierlings were in great demand as infantry support weapons due to their high rate of fire. The '88s' had an effective ceiling of 7900m (25,918ft) and were a good defence against the relatively rare, high altitude raids carried out by the Red Air Force. In the field, the gun's armour-piercing performance often led to AA batteries being deployed as anti-tank guns, which weakened the divisional air defence capability.

an increased rate of fire of 220rpm and a slightly reduced weight of 420kg (926lb).

Even as the FlaK 38 was entering service, the *Luftwaffe* and Army had doubts about its effectiveness, given the ever-increasing speeds of fighter-bombers and attack aircraft. The Army in particular felt the proper solution was to withdraw them from service and to standardize on the 37mm (1.45in) weapons they had been developing since the 1920s, which had a rate of fire about the same as the FlaK 38, but fired a round with almost eight times the volume which dramatically increased the effectiveness of each hit on the target. The 37mm (1.45in) weapons also offered significant advantages in terms of greater range and longer engagement times.

The 20mm (0.79in) class of AA weapons made a dramatic comeback in late 1940, when the first quadruple 2cm Flakvierling 38 were issued for combat trials. The compact design featured collapsing seats, folding control handles and ammunition racks. The mount had a triangular base with a jack at each leg for levelling the gun, which was fired by a pair of foot pedals. Each pedal

fired two diagonally opposite Flak 38s, with the option of automatic or semi-automatic fire.

Each of the four guns was fed by a 20-round magazine, giving a theoretical maximum combined rate of fire of 1400rpm. (The practical rate of fire was about 800rpm.) Its effective AA range was 2200m (7200ft). The towed version of the gun was normally carried on a two-wheeled Sd. Ah. 52 trailer, and could be towed by a variety of prime movers, such as the Opel Blitz, SdKfz 251 and SdKfz 11. Numerous self-propelled variants were also produced, the majority on halftracks such as the SdKfz 7/1.

FLAK TACTICS

The following notes are a summary of a 1943 US Army intelligence report on the tactical employment of flak units:

'The primary roles of flak units are:
■ Defence against hostile aerial reconnaissance.
■ Defence against hostile artillery observation.
■ Defence against air attacks on personnel and important installations.
■ Support of friendly air combat strength.

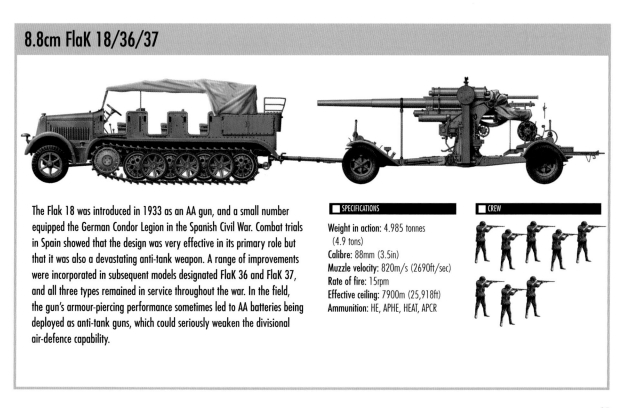

8.8cm FlaK 18/36/37

The Flak 18 was introduced in 1933 as an AA gun, and a small number equipped the German Condor Legion in the Spanish Civil War. Combat trials in Spain showed that the design was very effective in its primary role but that it was also a devastating anti-tank weapon. A range of improvements were incorporated in subsequent models designated FlaK 36 and FlaK 37, and all three types remained in service throughout the war. In the field, the gun's armour-piercing performance sometimes led to AA batteries being deployed as anti-tank guns, which could seriously weaken the divisional air-defence capability.

■ SPECIFICATIONS

Weight in action: 4.985 tonnes (4.9 tons)
Calibre: 88mm (3.5in)
Muzzle velocity: 820m/s (2690ft/sec)
Rate of fire: 15rpm
Effective ceiling: 7900m (25,918ft)
Ammunition: HE, APHE, HEAT, APCR

■ CREW

'Light, medium, and heavy AA weapons supplement each other in their effect. While the light and medium guns provide protection against low-flying enemy aircraft, the main role of heavy AA guns is to protect ground forces against air reconnaissance and high-altitude attacks while on the march, at rest, or in combat.

'In the battle for air superiority flak units are employed in their primary AA role. As control of the air is achieved, they become available for other missions, primarily infantry support and anti-tank duties. Whilst deployed in their AA role, flak guns are only permitted to fire on AFVs or other ground targets in self-defence. (As flak units are routinely deployed to protect field artillery and anti-tank batteries, they often find themselves in forward areas, and their very existence frequently depends on the ability to engage enemy AFVs.)

'Every German combat unit is responsible for its own low-altitude air defence – all riflemen are trained to fire on low-flying enemy aircraft as it has been found that concentrated rifle fire is effective up to slant ranges of about 1500ft. The effective AA ranges of other weapons are assessed as follows:

■ Machine gun fire: 2500ft.
■ 20mm and 37mm guns: 6000ft.

■ 88mm guns: 27000ft; however, their fire is regarded as ineffective against aircraft at altitudes below 1200ft – for this reason each heavy battery is supported by three 20mm Flakvierling, which are an organic part of the battery organization.

'Whilst on the move, panzer and panzergrenadier divisions are well spread out and where possible each division marches in several columns along parallel roads. The majority of AA guns are well forward in the columns and all defiles, bridges and stopping places are heavily defended by flak units. The columns usually halt after 2 hours for 20 minutes' rest, and after 4 or 5 hours' movement a halt of at least 3 hours is normal.

'In the event of air attack, the column continues its march under cover of fire from its light and medium flak which is supplemented by all the machine guns which can be brought to bear on the attacking aircraft.

'If the air attack proves to be so intense that severe casualties are likely to be inflicted on the truck-borne troops, the column halts and the troops take cover. The drivers, however, remain with their vehicles. When air attack threatens in open country, the tank columns deploy in open, usually V-shaped, formations.

2cm Flakvierling 38 (Sf) auf Fgst Zgkw 8t SdKfz 7/1

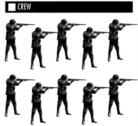

■ CREW

■ SPECIFICATIONS

Weight: 11.55 tonnes (11.36 tons)
Length: 6.85m (22ft 6in)
Width: 2.5m (8ft 2.5in)
Height: 3.3m (10ft 10in)
Engine: 104.4kW (140hp) Maybach HL 62 TUK 6cylinder petrol
Speed: 50km/h (31mph)
Range: 250km (155 miles)
Armament: 4 x 2cm (0.78in) Flakvierling 38 (sf)

The Germans were quick to appreciate the advantages of self-propelled light and medium AA guns, and the SdKfz 7 was one of the first vehicles to be adapted to carry the Flakvierling.

'The 88's potential as an anti-tank weapon was noted in the Spanish Civil War and its value in this role was confirmed in 1940/41 when it proved to be capable of destroying Allied tanks such as the Matilda, Char B and KV-1 which were invulnerable to the standard German 37mm anti-tank guns. It is now normal for a proportion of the available '88s' be assigned to anti-tank duties. The '88' has also been used in the infantry support role when it has proved invaluable in silencing heavily protected

Light Battery

Kübelwagen: 7	SdKfz 7: 1
Motorcyle: 12	Kfz 17/2 radio van: 5
Maultier: 3	SdKfz 7/1 SP Flakvierling: 12

The battery's SdKfz 7/1 were highly mobile and formidable weapons, both in their primary role of low altitude air defence and when deployed as infantry support weapons. Each vehicle's rate of fire was approximately 800rpm and this hail of 20mm (0.79in) HE and AP shells proved to be equally effective against ground and air targets.

97

bunkers by close range direct fire. It has also acted as field artillery in which role its high-velocity, flat trajectory fire has been highly effective.

In many cases, the '88s' have employed direct fire against AFVs and other ground targets – their excellent ZF 20-E telescopic sights are calibrated out to 9500 metres, although direct fire is highly unlikely at such ranges. When acting as field artillery, they are most likely to employ indirect fire controlled by an observation post.

Although the 20mm and 37mm flak guns lack the armour-piercing performance of the '88', their mobility and high rate of fire make them potent weapons against light AFVs. They are also extensively used against other ground targets, including bunkers which have been neutralized by concentrated fire against the weapons embrasures and observation ports. In common with the '88s', they have also given covering fire to engineers engaged in bridging operations. Only direct fire is employed using simple telescopic sights, in conjunction with tracer ammunition.'

A *Das Reich* '88' flak gun employed in the tank detroyer role deploys in a wheat field alongside a Panzer IIIN somewhere in the Kursk region. The '88' proved to be a highly adaptable and effective weapon.

2nd SS Artillery Regiment

By the time of the Kursk offensive, *Das Reich*'s artillery regiment was equipped with a mix of towed and self-propelled 105mm (4.1in) and 150mm (5.9in) howitzers.

The 10.5cm leFH 18 was designed and developed by Rheinmetall in 1929–30 and it entered service in 1935, becoming the standard German divisional field howitzer of World War II. As the German automotive industry would be incapable of producing sufficient vehicles to fully motorize the army for many years, the design had to allow for both horse and motor traction.

TOWED ARTILLERY:
10.5CM LEFH 18 AND 10.5CM LEFH 18M

The horse-drawn versions that equipped infantry divisions were fitted with wooden-spoked wheels, whilst those destined for mechanized formations had pressed steel wheels with solid rubber tyres. It had a simple sliding block breech mechanism with a hydro-pneumatic recoil system. The type proved to be an effective weapon in the campaigns of 1939/40, but was outranged by Allied field artillery such as the British 25 pdr and the French 75mm (2.95in) and 120mm (4.7in) guns. Unsurprisingly, the gunners demanded an improvement to its maximum range of 10,675m (35,020ft) and Rheinmetall began a series of studies to resolve the problem.

By 1941, trials had shown that the addition of a muzzle brake would allow the use of larger propelling charges to achieve greater range. The recoil system was modified to cope with the more powerful charges and the new howitzer was approved for service as the 10.5cm leFH 18M; the 'M' was an abbreviation of *Mundungbremse* – muzzle brake. Long-range shells were also produced to maximize the new design's potential, the first of which was the FH Gr Fern, which increased the howitzer's maximum range to 12,325m (40,440ft). Whilst this was a marked improvement, it was felt that more could be achieved and a variety of experimental ammunition types were produced. The single-baffle muzzle brake fitted to early production weapons was found to have relatively low efficiency and was later improved by welding two protruding ears to the rear of the port. However, this solution upset the ballistics of early types of experimental long-range fin-stabilized and discarding-sabot shells and an entirely new cage-type muzzle brake replaced the earlier types. (Ironically, all the effort that went into redesigning the muzzle brake was largely wasted, as very few of

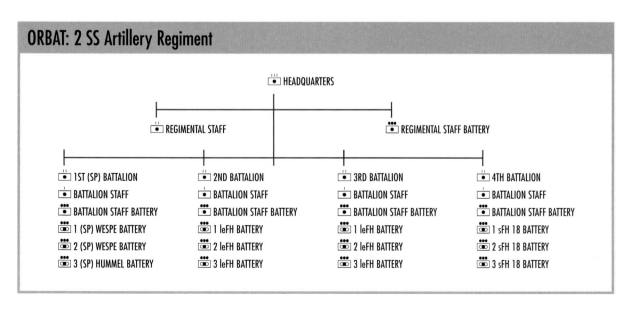

ORBAT: 2 SS Artillery Regiment

- HEADQUARTERS
- REGIMENTAL STAFF
- REGIMENTAL STAFF BATTERY

1ST (SP) BATTALION	2ND BATTALION	3RD BATTALION	4TH BATTALION
BATTALION STAFF	BATTALION STAFF	BATTALION STAFF	BATTALION STAFF
BATTALION STAFF BATTERY	BATTALION STAFF BATTERY	BATTALION STAFF BATTERY	BATTALION STAFF BATTERY
1 (SP) WESPE BATTERY	1 leFH BATTERY	1 leFH BATTERY	1 sFH 18 BATTERY
2 (SP) WESPE BATTERY	2 leFH BATTERY	2 leFH BATTERY	2 sFH 18 BATTERY
3 (SP) HUMMEL BATTERY	3 leFH BATTERY	3 leFH BATTERY	3 sFH 18 BATTERY

the many versions of experimental long-range shells were approved for service use and fewer still ever reached frontline units.)

15CM SFH 18

The howitzer originated with several competing designs from Rheinmetall and Krupp, all of which were deemed unsatisfactory. However, the requirement for a new 150mm (5.9in) howitzer was becoming increasingly urgent and it was decided to combine the best features of both designs, by adopting the Rheinmetall gun on a Krupp carriage, which entered service in May 1935.

The carriage was a relatively standard split-trail design carrying earth spades, which were fitted to the ends of the trails to give added stability when firing. In view of the slow progress in motorizing the German army, the howitzer was primarily designed for horse traction, with an unsprung axle and solid rubber tyres. A two-wheeled bogie was introduced to allow the sFH 18 to be towed by motor vehicles, but the lack of suspension severely restricted the maximum towing speed. By the outbreak of war, over 1300 of these howitzers had been completed and production continued throughout the war, with a total of 2295 in service in 1944.

In action in Russia, the sFH 18 proved to be greatly inferior to the Red Army's 122mm (4.8in) A-19 gun and 152mm (6in) ML-20 gun-howitzer, both of which outranged it by a substantial margin. This led to numerous efforts to develop entirely new designs to out-perform the Soviet artillery, whilst various experiments were also carried out on the sFH 18 to improve its range. These led to the development of the 15cm sFH 18M with a removable barrel liner and a muzzle brake that allowed the use of a more powerful propelling charge. The 18M's maximum range was increased to 15,100m (49,500ft), but it was found that the barrels suffered increased wear and the recoil system could not handle the increased loads despite the muzzle brake. This led to the introduction of the 15cm R. Gr. 19 FES, a rocket-assisted round that could reach 18,200m (59,710ft), which reduced the range advantage of the Red Army's artillery but was significantly less accurate than conventional ammunition.

Several further developments of the sFH 18 were produced. The 15cm sFH 36 incorporated many

light alloy components, which reduced its weight to 3450kg (7600lb) in an attempt to improve mobility. However, the increased cost and the need to conserve light alloys for aircraft production led to it being rejected for service use. The 15cm sFH 40 was another improved version, featuring a slightly longer barrel and a new carriage that brought the range up to 15,400m (50,520ft) with conventional ammunition. However, this type was even heavier than the sFH 18 and was found to be too awkward for service use, although some of these barrels were later fitted to existing sFH 18 carriages, creating a hybrid version designated sFH 18/40.

SELF-PROPELLED ARTILLERY: WESPE

The opening stages of Operation *Barbarossa* showed that the Panzer II was thoroughly obsolete as a light tank, and in early 1942 it was formally relegated to second-line duties. This decision allowed a number of hulls to be made available for conversion to self-propelled guns and a prototype Wespe (Wasp) was built mounting the 10.5cm leFH 18/2 howitzer. This underwent trials alongside similar designs based on the Panzer III and Panzer IV, as a result of which the Wespe was approved for service.

The initial production models were based on the Panzer II Ausf F with the engine moved to the centre of the hull and a reinforced suspension to absorb the howitzer's recoil. Later examples were based on a modified chassis with a slightly lengthened hull, and the suspension was again strengthened. All vehicles carried their main armament on a limited traverse mount in an open-topped fighting compartment which provided very cramped accommodation for the commander and three-man gun crew. The howitzer had a total traverse of 34° and elevation of −5° to +42° – all types of 105mm (4.1in) ammunition could be fired to a maximum range to 12,325m (40,440ft). The lack of space meant that only 32 rounds could be stowed in the fighting compartment and an ammunition carrier variant was designed to overcome this problem. The vehicle was simply a standard Wespe with its main armament removed and stowage racks fitted for a total of 90 rounds of 105mm (4.1in) ammunition.

The Wespe was a quick and easy way of using an obsolete tank hull to produce an effective self-

1st (Self-Propelled) Battalion

Kübelwagen: 14   Kfz 17 radio van: 6   Artillerie-Panzerbeobachtungswagen III: 6
Motorcycle: 1   Schwimmwagen: 3   SdKfz 251/18: 3
Kettenkrad 4   Maultier: 1   Wespe SP gun: 8   Hummel SP gun: 8

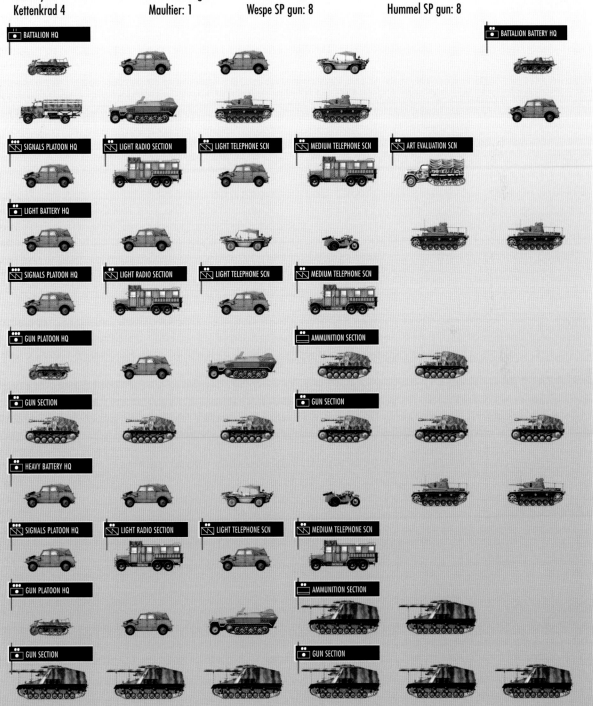

propelled howitzer. Despite its limitations, a total of 676 (including ammunition carriers) were produced during the period from February 1943 to August 1944, and the type remained in service throughout the war.

HUMMEL

Development of the Hummel began in July 1942 and initial studies concentrated on mounting a 105mm (4.1in) howitzer on the hull of the Panzer III or Panzer IV before it was realized that it should be feasible to improve the vehicle's firepower by fitting a 150mm (5.9in) howitzer. The solution finally adopted was to mount the 15cm sFH 18 L/30 howitzer on the specially designed Geschützwagen III/IV, which used elements of both the Panzer III (driving and steering system) and Panzer IV chassis (suspension and engine).

The Hummel carried its main armament on a limited traverse mount in an open-topped, thinly armoured five-man fighting compartment. (The driver was isolated in a separate enclosed position in the hull front, but late production models had a redesigned front superstructure, with fully-enclosed positions for the radio operator and driver. This gave welcome additional space in the fighting compartment, which now only had to accommodate the commander and three gunners.)

As the Hummel could carry only 18 rounds of main armament ammunition, the *Munitionsträger Hummel* (ammunition carrier Hummel) was developed. This was a standard production vehicle without the howitzer (a 10mm/0.4in plate was fitted to cover the gun mount) but fitted with racks for 40 rounds of 150mm (5.9in) ammunition.

The type proved to be highly successful and remained in service throughout the war. Total production amounted to 714 vehicles plus a further 150 ammunition carriers.

ARTILLERY TACTICS

In common with the general German practice of *Auftragstaktik* (mission command), all artillery orders were initially given verbally and only later confirmed in writing. Whenever possible, these verbal orders were given in a briefing session held on ground that offered a clear view of the target area

rather than simply referring to the map. The divisional artillery commander's order was not issued as an annex to the division order, but as a separate artillery order.

German artillery tactics were generally simple but effective, with a few standard techniques being adopted, with variations kept to a bare minimum. The aim was to ensure a rapid response to requests for artillery support by keeping things simple, rather than diverting resources to the preparation of several highly refined techniques, each suitable for only a few complex situations. This standardization of technique was in contrast to general German tactical doctrine, which stressed the necessity of working out an independent solution to a specific problem rather than applying a rigid prearranged formula.

Concentrated fire was always stressed as the ideal; whenever possible, at least a full battalion would fire on any given target. Only in exceptional circumstances – such as when its parent division was covering a very large sector of the front – would the battalion be broken down into individual batteries. Such concentrated fire was especially important in counter-battery missions, when the suppression of enemy artillery fire had to be achieved quickly to prevent heavy damage and casualties. Ideally, medium artillery would be primarily responsible for counter-battery fire because it had a longer range than field artillery and its heavier shells were more lethal against enemy artillery positions.

Whilst radio links were set up between observation posts and gun positions, other communications were normally by field telephone, except when on the move. This improved security, but meant a heavy workload for the signals staff who had to lay and maintain miles of telephone lines connecting the various elements of each unit.

In preparatory bombardments preceding German attacks, each battery normally covered one or more targets, each about 100–150m (328–492ft) in width. Considerable emphasis was placed on flexibility in fire planning and procedures, particularly by irregular bursts of fire targeting enemy infantry, artillery and command posts, as well as the sector of front to be attacked.

Although veteran gun crews could achieve

15cm schwere Feldhaubitze 18 (15cm sFH 18)

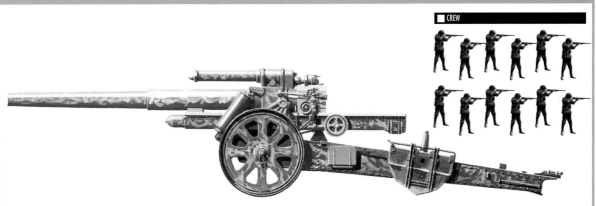

■ CREW

This howitzer was an unusual weapon in that the ordnance was a Rheinmetall design whilst the carriage was designed by Krupp. It entered service in 1935 and was regarded as highly satisfactory until 1941, when it became clear that it was outranged by its Soviet counterparts. Various attempts were made to improve the howitzer's range, including more powerful propellant charges, and one of the earliest rocket-assisted shells, the 15cm R. Gr. 19 FES. These provided only a partial solution to the problem and entirely new 150mm (5.9in) weapons were proposed. However, none of these entered production and the 15cm sFH 18 remained in service throughout the war.

■ SPECIFICATIONS

Weight in action: 5.53 tonnes (5.44 tons)
Calibre: 149.1mm (5.89in)
Muzzle velocity: 495m/sec (1624ft/sec)
Range: (HE) 13,250m (43,470ft) (Rocket Assisted HE): 18,200m (59,710ft)
Ammunition: HE, Rocket-Assisted HE, HEAT,
Shell weight: (HE) 43.5kg (95.9lb)

10.5cm leichte FeldHaubitze 18M (10.5cm leFH 18M)

■ CREW

The 10.5cm leFH 18M gradually superseded the 10.5cm leFH 18 as the standard German divisional field howitzer from early 1943. It was designed by Rheinmetall in 1940/41 in an effort to get more range from the basic leFH 18 design. A muzzle brake was fitted and the recoil system adjusted to allow the use of a more powerful charge and new long-range shell. These measures increased its range from 10,675m to 12,325m (35,020ft to 40,440ft), but the dire shortage of light alloys meant that nothing could be done to reduce the howitzer's excessive weight.

■ SPECIFICATIONS

Weight in action: 1.985 tonnes (1.95 tons)
Calibre: 105mm (4.13in)
Muzzle velocity: 540m/sec (1772ft/sec)
Range: (HE) 12,325m (40,440ft)
Ammunition: HE, HEAT, Smoke
Shell weight: (HE) 14.8kg (32.63lb)

exceptionally high rates of fire for a few minutes at a time, the following table shows the maximum rates of fire when following basic safety drills:

Howitzer Rate of Fire

WEAPON	SHORT BURST	RPM	PROLONGED FIRE
105mm howitzer	6	2	1/2
150mm howitzer	4	1	1/3

The Germans would frequently detach a single gun, deploying it far enough from the rest of the battery to appear to be a different position. This was used for harassing fire and fire against high-priority targets, and to deceive enemy observers.

Much effort went into ensuring that infantry commanders understood the characteristics, capabilities and limitations of artillery in order to secure the most effective support. Their training emphasized that:

■ The effectiveness of artillery depends on the neutralization of enemy artillery; consequently some of the fire must be employed on counter-battery missions.

■ The ammunition supply is limited, and the laying of heavy concentrations on important areas means a loss of fire on less important ones.

■ The artillery should engage only those targets which justify its heavy fire.

■ Unnecessary or too hasty requests for fire support would divert artillery from its principal missions.

One factor that ensured infantry commanders were familiar with their artillery's capabilities and limitations was the presence of the infantry gun company in the infantry regiment. This had several benefits; it relieved the artillery from many small but difficult direct-support missions, allowing it concentrate on its primary roles. Most importantly, it lessened the artillery-infantry gap, so that liaison was not between two distinct and separate units of

SdKfz 165 Hummel

■ CREW

■ SPECIFICATIONS

Weight: 24.384 tonnes (24 tons)
Length: 7.17m (23ft 6in)
Width: 2.87m (9ft 5in)
Height: 2.81m (9ft 3in)
Engine: 98Kw (265hp) Maybach HL 120 TRM V12 petrol
Speed: 42km/h (26mph)
Range: 215km (134 miles)
Armament: 1 x 150mm (5.9in) sFH18/1 L/30 howitzer, plus 1 x 7.92mm (0.31in) MG34 or MG42 machine gun

The Hummel was armed with the 15cm sFH 18 L/30 howitzer on the specially designed hull designated Geschützwagen III/IV, which combined elements of both the Panzer III (driving and steering system) and Panzer IV chassis (suspension and engine). The Kursk offensive was the type's first major action, in which they served in the artillery battalions of panzer and panzer grenadier divisions. They were deployed in separate heavy self-propelled artillery batteries, each with 6 Hummels and one ammunition carrier. (The latter was modified by the removal of the main armament and the addition of stowage for 40 rounds of 150mm ammunition.) Production ran from late 1942 until 1944, by which time 714 Hummels and 150 ammunition carriers had been completed. The type remained in service throughout the war.

artillery and infantry, but rather between the regular artillery of the supporting battalion and an infantry unit that already had organic artillery weapons. This allowed the coordination of fire plans as well as mutual observation by the infantry gun company and the artillery.

Infantry regimental and battalion commanders were trained to assist the commander of their supporting artillery by continually updating him regarding the infantry plan of action, its progress and its need for artillery support. One of the infantry's primary tasks was to seize and hold the forward areas required for artillery observation in order to maximize the effectiveness of their artillery support. It was also stressed that the infantry should be informed of the positions of the artillery forward observers, observation posts and command posts.

Liaison between the two arms was ensured by attaching forward artillery observers to work with the leading infantry units. These forward observers with man-pack radios were regarded as the only effective means of obtaining satisfactory observation of potential targets. Radios were used both by individual

batteries and by battalions and it was not uncommon for the battery commander himself to act as a forward observer, particularly at the beginning of an engagement when he was unfamiliar with the terrain.

The short-range artillery firepower of formations such as *Das Reich* was formidable, especially as they deployed infantry guns on a far greater scale than Allied armies. However, there were distinct weaknesses in long-range fire support capabilities. All the German 105mm (4.1in) howitzers to see service in large numbers were outranged by even the most elderly Soviet 76.2mm (3in) divisional guns. In terms of medium artillery the situation was just as bad – the various versions of German 150mm sFH 18 howitzer were all outranged by the Red Army's 122mm (5.9in) A-19 guns and 152mm (6in) ML-20 gun/howitzers. (The introduction of the German R. Gr. 19 FES rocket-assisted ammunition was no more than a partial solution to the problem – it outranged the ML-20, but still had a shorter range than the A-19. It was issued in only small quantities and its usefulness was further limited by its poor accuracy and small HE payload.)

SdKfz 124 Wespe

■ SPECIFICATIONS

Weight: 11.176 tonnes (11 tons)
Length: 4.82m (15ft 10in)
Width: 2.28m (7ft 6in)
Height: 2.31m (7ft 7in)
Engine: 104kW (140hp) Maybach HL62TR 6-cylinder petrol
Speed: 40km/h (25mph)
Range: 140km (87 miles)
Armament: 1 x 105mm (4.1in) le FH18/2 L/26 howitzer, plus 1 x 7.92mm (0.31in) MG34 machine gun

The Wespe entered service with the artillery battalions of panzer and panzergrenadier divisions in spring 1943. Each battery had six Wespe and two more acting as ammunition resupply vehicles. (These were modified by the removal of the main armament and the addition of stowage for 90 rounds of 105mm/4.1in ammunition). The type's first major action was at Kursk, where it proved to be highly effective, and it remained in service on all fronts until the end of the war. Production continued until mid-1944, by which time 682 had been built and a further 158 completed as ammunition carriers.

Panzerjäger Battalion

Even in the pre-war era, German anti-tank units were given considerable importance, with an emphasis on their offensive potential rather than a purely defensive role.

This offensive doctrine was formalized in April 1940, when all *Panzerabwehr* (anti-tank) units were redesignated as *Panzerjäger* (tank hunter) units to coincide with the introduction of the first German self-propelled anti-tank gun, the Panzerjäger I. This used the hull of the Panzer I to carry the Czech 47mm (1.85in) anti-tank gun in a thinly armoured, open-topped fighting compartment. The Panzerjäger I could deal with most British and French tanks of 1940 (although its gun was only marginally effective against the Char B and Matilda at close range) but it was outclassed by the Red Army's T-34s and KV-1s. Despite this, the Panzerjäger concept had proved its worth – obsolete tank hulls could be used to give anti-tank guns a far higher degree of mobility than their towed equivalents and could be produced far more quickly and cheaply than conventional tanks or assault guns.

The urgent need for heavier and more mobile anti-tank firepower to counter Soviet armoured forces led to a major expansion of Panzerjäger units in 1942. Several obsolete and obsolescent AFVs were used as the basis for new Panzerjägers, including captured French vehicles such as the Lorraine (Marder I), the Panzer II (Marder II) and the Panzer 38(t) (Marder III). All these followed a similar basic design incorporating: an open-topped, thinly armoured fighting compartment; a limited traverse main armament comprising either a captured Soviet 76.2mm (3in) gun (sometimes modified to fire German 75mm/2.95in ammunition) or a 75mm (2.95in) PaK 40 anti-tank gun.

By mid-1943, favoured formations such as *Das Reich* were increasingly being equipped with the Marder III Ausf M (*Mittelmotor* – middle engine). This was the final version of the Marder III and was armed with the 75mm (2.95in) PaK 40 anti-tank gun. In this variant, the engine was moved from the rear to a central position between driver and the rest of the crew. This allowed the fighting compartment to be moved to the rear and lowered to use the full depth of the former engine compartment. The change improved battlefield survivability by decreasing the vehicle's height, making it less conspicuous than earlier versions.

The armour of the open-topped fighting compartment was redesigned to incorporate a rear plate, and offered the gun crew far better protection from small arms fire and shell splinters than previous Marders. In these earlier vehicles, the commander was badly overworked because he also acted as the gunner. The Ausf M significantly improved combat efficiency by moving the radio operator from his former position alongside the driver to the fighting compartment, where he now also acted as loader for the main armament. This freed the commander from his previous gunnery duties and allowed him to concentrate on command responsibilities.

In common with all other Panzerjäger, the Marder III was thinly armoured and required careful handling to survive on the battlefield, but its Praga engine was highly reliable and its firepower was adequate to deal with the majority of Soviet AFVs. More than 1700 Marder IIIs were produced

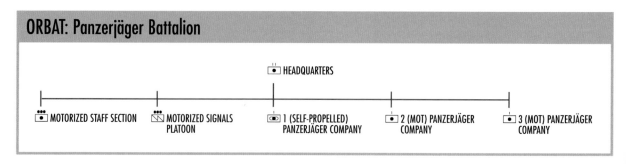

ORBAT: Panzerjäger Battalion

HEADQUARTERS

MOTORIZED STAFF SECTION | MOTORIZED SIGNALS PLATOON | 1 (SELF-PROPELLED) PANZERJÄGER COMPANY | 2 (MOT) PANZERJÄGER COMPANY | 3 (MOT) PANZERJÄGER COMPANY

between 1942 and early 1944, remaining in service throughout the war.

PAK 40

There were never enough Panzerjäger to meet the need for tank destroyers and even elite units such as *Das Reich* had to rely on towed anti-tank guns to bring the number of anti-tank weapons up to the required level.

In 1939 the German army issued a specification for a 75mm (2.95in) anti-tank gun (*Panzerabwehrkanone*) to Krupp and Rheinmetall, but its development was a low priority until the opening stages of Operation *Barbarossa* revealed the virtual invulnerability of the T-34 and KV-1 to all standard German anti-tank weapons. The PaK 40 was immediately subjected to a crash development programme, but until it became available in quantity a variety of stop-gap weapons were fielded, including more than 1000 captured Soviet 76.2mm (3in) F-22 divisional guns, which were converted to become highly effective anti-tank guns under the designation PaK 36(r).

Panzerjäger Battalion: Equipment

UNIT	EQUIPMENT	STRENGTH	UNIT	EQUIPMENT	STRENGTH
Battalion HQ	Kübelwagen	2	1 Motorized Anti-Tank Company		
	Kettenkrad	2	Company HQ	Kettenkrad	1
	SdKfz 251/8	1		Kübelwagen	2
AA Platoon			1 Platoon		
Platoon HQ	Motorcycle	1	Platoon HQ	Kettenkrad	1
	Kübelwagen	1		Kübelwagen	1
1 Section	SdKfz 10/5	1	1 Section	Maultiers or SdKfz 11	2
	Maultier	1		75mm Pak 40	2
2 Section	SdKfz 10/5	1	2 Section	Maultier or SdKfz 11	2
	Maultier	1		75mm Pak 40	2
3 Section	SdKfz 10/5	1	2 Platoon		
	Maultier	1	Platoon HQ	Kettenkrad	1
Signals Platoon				Kübelwagen	1
Platoon HQ	Kübelwagen	2	1 Section	Maultier or SdKfz 11	2
Telephone Section	Kfz 17/2	1		75mm Pak 40	2
1 Portable Radio Section	Kübelwagen	1	2 Section	Maultier or SdKfz 11	2
2 Portable Radio Section	Kübelwagen	1		75mm Pak 40	2
1 Light Radio Section	Kfz 17/1 radio car	1	2 Motorized Anti-Tank Company		
2 Light Radio Section	Kfz 17/1 radio car	1	Company HQ	Kettenkrad	1
Self-Propelled Anti-Tank Company				Kübelwagen	2
Company HQ	Motorcycle	1	1 Platoon		
	Kfz 15 radio car	1	Platoon HQ	Kettenkrad	1
	Marder III Ausf M	1		Kübelwagen	1
1 Platoon			1 Section	Maultier or SdKfz 11	2
HQ	Kübelwagen	1		75mm Pak 40	2
	Motorcycle combination	1	2 Section	Maultier or SdKfz 11	2
Gun Section	Marder III Ausf M	3		75mm Pak 40	2
2 Platoon			2 Platoon		
HQ	Kübelwagen	1	Platoon HQ	Kettenkrad	1
	Motorcycle combination	1		Kübelwagen	1
Gun Section	Marder III Ausf M	3	1 Section	Maultier or SdKfz 11	2
3 Platoon				75mm Pak 40	2
HQ	Kübelwagen	1	2 Section	Maultier or SdKfz 11	2
	Motorcycle combination	1		75mm Pak 40	2
Gun Section	Marder III Ausf M	3			

Panzerjäger Battalion

Kübelwagen: 18	SdKfz 10/5 flak: 3	Maultier: 3	Marder III Ausf M: 10
Kettenkrad: 8	Kfz 15 radio car: 3	SdKfz 11/PaK 40: 16	
SdKfz 251/8: 1	Kfz 17 radio car: 1	Motorcycle: 5	

BATTALION HQ

AA PLATOON HQ

AA SECTION   AA SECTION   AA SECTION

SIGNALS PLATOON HQ   TELEPHONE SECTION

PORTABLE RADIO SECTION   PORTABLE RADIO SECTION   LIGHT RADIO SECTION   LIGHT RADIO SECTION

1 SELF-PROP AT COY HQ

1 AT PLATOON HQ   AT GUN SECTION

2 AT PLATOON HQ   AT GUN SECTION

3 AT PLATOON HQ   AT GUN SECTION

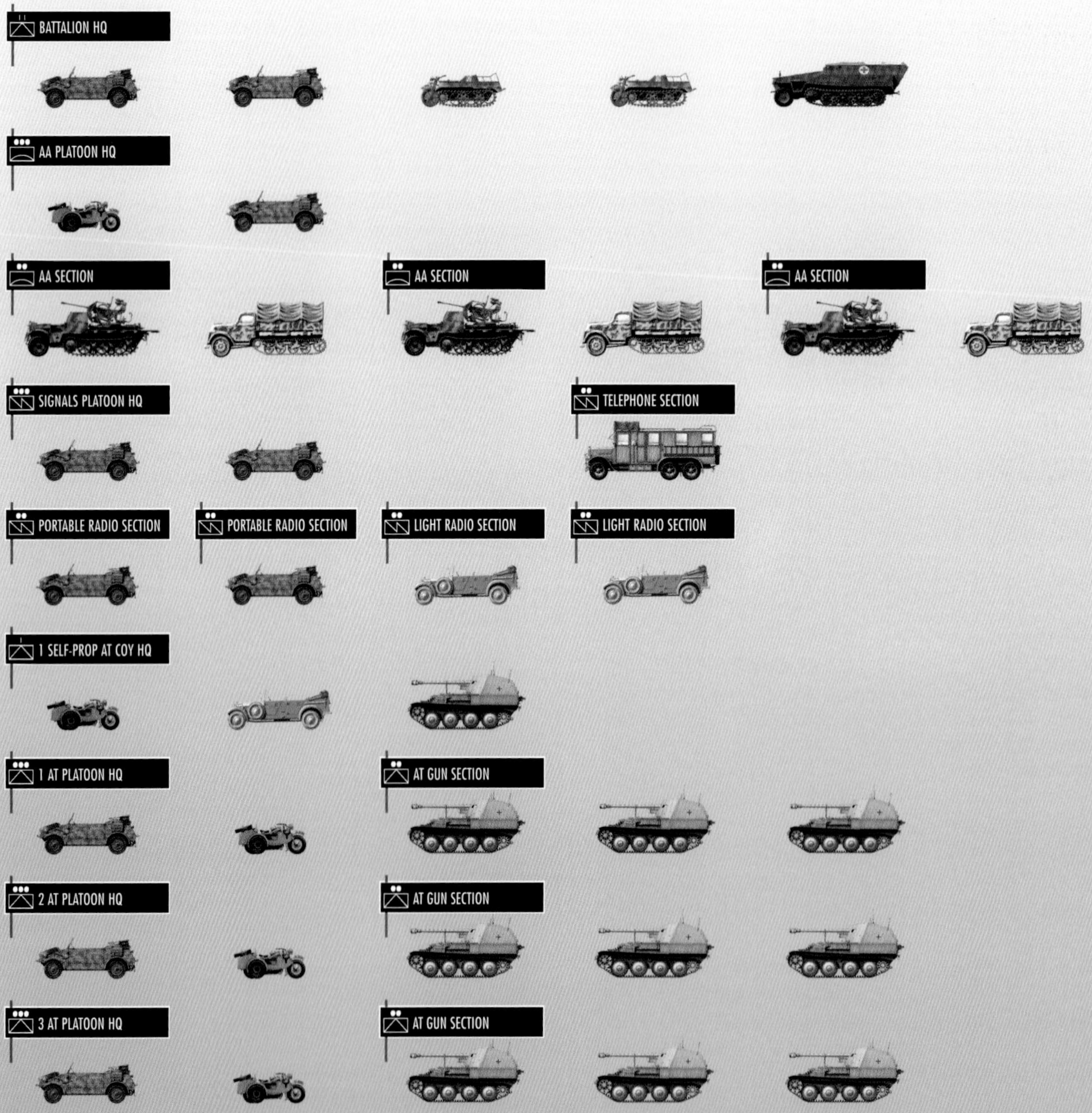

German Army notes issued around the time of the Kursk offensive stated that the main task of Panzerjägers in defence was the destruction of tanks that had broken through. They were therefore held as a mobile reserve and employed en masse at the point of main effort.

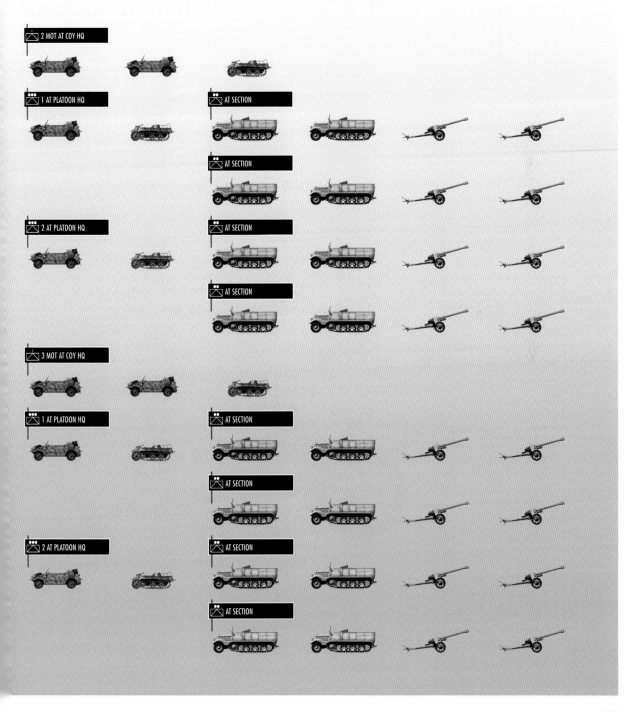

2 MOT AT COY HQ

1 AT PLATOON HQ

AT SECTION

AT SECTION

2 AT PLATOON HQ

AT SECTION

AT SECTION

3 MOT AT COY HQ

1 AT PLATOON HQ

AT SECTION

AT SECTION

2 AT PLATOON HQ

AT SECTION

AT SECTION

A few early production PaK 40s may have reached the front line in late 1941, but it was several months before it was in service in significant numbers. Although the gun was an accurate and powerful weapon, it was heavy and almost impossible to manhandle in the thick mud characteristic of the Russian spring and autumn. A large number of the guns lost on the Eastern Front were simply abandoned by retreating forces after becoming hopelessly bogged down.

ANTI-TANK TACTICS

By mid-1943, two years of hard-won combat experience against the Red Army led to the issue of a series of tactical notes on the handling of anti-tank units. These notes featured a wide range of instructions, including:

■ The tendency to split up anti-tank units completely, in order to have a proportion of anti-tank firepower everywhere, is wrong. The smallest permissible unit is the section (two guns), except when defending streets where single guns may be deployed.

■ Whole companies, or at least full platoons, should cover likely tank approaches. To use a single anti-tank gun is to invite destruction.

■ Engagement of even worthwhile infantry targets must be the exception rather than the rule. Such employment is limited by the gun's lack of mobility, by its vulnerability and by the limited issue of high-explosive shells.

■ In the attack, anti-tank units follow the advancing infantry in areas likely to favour tank counter-attacks, moving from cover to cover in such a manner that their guns always have advantageous positions. The leading infantry must not be beyond their range and as many guns as possible must be ready to fire simultaneously. Close liaison between anti-tank units and the infantry is essential. When the objective has been reached, or if the attack is held up, a solid defensive belt must be organized immediately by the anti-tank unit commander.

■ In defensive operations, an anti-tank defence plan will be drawn up by the responsible anti-tank commander. Location of the main defensive belt must give his guns suitable fields of fire.

■ Anti-tank positions must be established at some distance to the rear and must be camouflaged so they will not be seen and subjected to bombardment before the attack. However, when

7.5cm Panzerabwehrkanone 40 (7.5cm PaK 40)

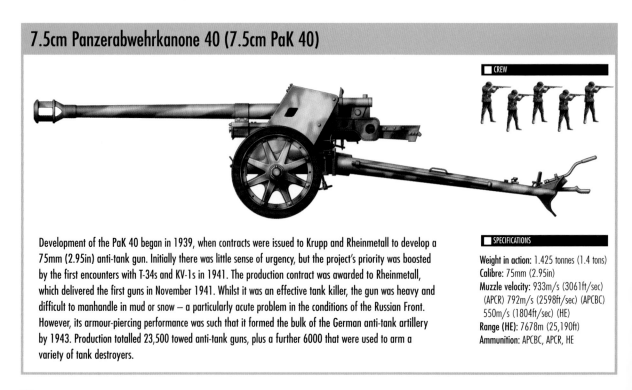

■ CREW

Development of the PaK 40 began in 1939, when contracts were issued to Krupp and Rheinmetall to develop a 75mm (2.95in) anti-tank gun. Initially there was little sense of urgency, but the project's priority was boosted by the first encounters with T-34s and KV-1s in 1941. The production contract was awarded to Rheinmetall, which delivered the first guns in November 1941. Whilst it was an effective tank killer, the gun was heavy and difficult to manhandle in mud or snow – a particularly acute problem in the conditions of the Russian Front. However, its armour-piercing performance was such that it formed the bulk of the German anti-tank artillery by 1943. Production totalled 23,500 towed anti-tank guns, plus a further 6000 that were used to arm a variety of tank destroyers.

■ SPECIFICATIONS

Weight in action: 1.425 tonnes (1.4 tons)
Calibre: 75mm (2.95in)
Muzzle velocity: 933m/s (3061ft/sec)
(APCR) 792m/s (2598ft/sec) (APCBC)
550m/s (1804ft/sec) (HE)
Range (HE): 7678m (25,190ft)
Ammunition: APCBC, APCR, HE

selecting positions it must be remembered that these should be sufficiently far forward to cover the ground in front of the main defensive belt.

■ Mines and obstacles should be used in suitable areas. Tank-hunting detachments should be held ready in villages, wooded areas and close country.

■ Nests of anti-tank guns should be established under a single unified command. Units arriving subsequently will be incorporated in the general anti-tank defence plan.

■ Open fire as late as possible. Do not be deceived by feint attacks. Use one uniform system of tank warning. Tank warnings have absolute priority.

In the employment of self-propelled anti-tank guns (Panzerjäger), the following will apply:

■ Companies are controlled by radio, in emergency by flag signals. Normal formations for movement on the battlefield are file, arrowhead, broad arrowhead, or extended line. Panzerjäger use fire and movement, their constant readiness for action making them the ideal mobile reserve. They are therefore, the weapon to use at points of main effort. The tactical unit is the company; exceptionally, the platoon. Panzerjäger can only be employed on open flanks if adequately covered by infantry. Whenever possible, ground reconnaissance, preferably on foot, must precede the occupation of positions.

■ On the move, one Panzerjäger half-platoon should be as far forward as possible; the remainder of the platoon should be with the advance party. The rest of the company will remain together. Road reconnaissance must include investigation of the carrying capacity of bridges.

■ In an attack, the infantry will be accompanied by Panzerjäger platoons, each giving the other mutual support. The enemy should be engaged by surprise, when possible from defiladed positions or from positions on reverse slopes, with all guns firing simultaneously. Whenever possible, fire should be by whole companies, since it will frequently be necessary to fire in several directions at the same time. Platoons can fire effectively only in one direction at a time.

■ The only completely successful method of employing Panzerjäger companies is in mobile operations. Flank attacks are very effective, especially if they are combined with a small frontal attack.

Marder III SdKfz 138 Ausf M

■ CREW

■ SPECIFICATIONS

Weight: 10.8 tonnes (10.63 tons)
Length: 5.77m (18ft 11in)
Width: 2.15m (7ft 1in)
Height: 2.5m (8ft 2in)
Engine: 112kW (150hp) Praga EPA 6-cylinder petrol
Speed: 35km/h (21.75mph)
Range: 240km (149 miles)
Armament: 1 x 75mm (2.95in) PaK40/3 gun, plus 1 x 7.92mm (0.31in) MG34 machine gun

This was the second Panzerjäger to be based on the hull of the reliable Panzer 38(t). Unlike the earlier conversion, it utilized the fighting compartment of the Panzer 38(t) to produce a lower silhouette and provide slightly better protection for the gun crew.

2nd Sturmgeschütz Abteilung

The Sturmgeschütz III (StuG III) was one of the most successful German AFVs of the entire war. A total of approximately 9500 vehicles were produced from January 1940 to March 1945 and as early as the spring of 1944 StuG III units claimed the destruction of no less than 20,000 Allied tanks.

The Sturmgeschütz III originated from German experiences in World War I, when it was discovered that attacking infantry lacked the means to effectively engage fortifications. The artillery of the time was too heavy and immobile to keep up with the advancing infantry to destroy bunkers, pillboxes and other minor obstacles with direct fire. In 1935 an initial study was drafted by the then Colonel Erich von Manstein and submitted to General Ludwig Beck, suggesting that *Sturmartillerie* ('assault artillery') units should be used in a direct-fire support role for infantry divisions.

The idea was favourably received and on 15 June 1936, Daimler-Benz AG received an order to develop an armoured infantry support vehicle capable of mounting a 75mm (2.95in) artillery piece. The gun mount's fixed, fully-integrated casemate superstructure was to allow a limited traverse of a minimum of 25° and provide overhead protection for the crew. The height of the vehicle was not to exceed that of the average man on foot.

Daimler-Benz used the chassis and running gear of its recently designed Panzer III medium tank as a basis for the new vehicle. Prototype manufacture was passed over to Alkett, which produced five examples in 1937 of the experimental 0-series StuG based upon the Panzer III Ausf. B. All these prototypes had a mild steel superstructure and Krupp's short-barrelled 75mm (2.95in) StuK 37 L/24 gun. This model was known as the Sturmgeschütz Ausführung A.

Sturmgeschütz III (StuG III) Ausf G

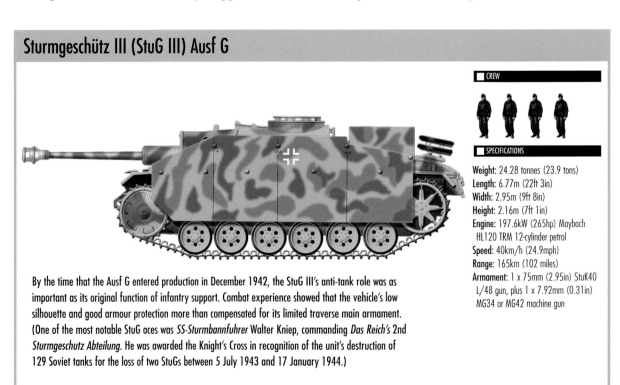

CREW

SPECIFICATIONS

Weight: 24.28 tonnes (23.9 tons)
Length: 6.77m (22ft 3in)
Width: 2.95m (9ft 8in)
Height: 2.16m (7ft 1in)
Engine: 197.6kW (265hp) Maybach HL120 TRM 12-cylinder petrol
Speed: 40km/h (24.9mph)
Range: 165km (102 miles)
Armament: 1 x 75mm (2.95in) StuK40 L/48 gun, plus 1 x 7.92mm (0.31in) MG34 or MG42 machine gun

By the time that the Ausf G entered production in December 1942, the StuG III's anti-tank role was as important as its original function of infantry support. Combat experience showed that the vehicle's low silhouette and good armour protection more than compensated for its limited traverse main armament. (One of the most notable StuG aces was *SS-Sturmbannfuhrer* Walter Kniep, commanding *Das Reich's* 2nd *Sturmgeschutz Abteilung*. He was awarded the Knight's Cross in recognition of the unit's destruction of 129 Soviet tanks for the loss of two StuGs between 5 July 1943 and 17 January 1944.)

Whilst the StuG III was considered self-propelled artillery, it was not clear which arm of service would handle the new weapon. The Panzer arm, which was the natural user of AFVs, had no resources to spare for the formation of StuG units, and neither did the infantry. After some discussion, it was agreed that the new type would best be employed as part of the artillery. StuGs were organized into battalions (later renamed 'brigades' for disinformation purposes) and followed their own specific doctrine, initially concentrating on infantry support using direct fire, but with increasing emphasis on the anti-tank role.

As the StuG III was designed for the close support role, early models were fitted with a low-velocity 75mm (2.95in) StuK37 L/24 gun to destroy soft-skin targets and field fortifications. After initial encounters with the Soviet KV-1 and T-34 tanks, the StuG III was re-equipped with a high-velocity 75mm (2.95in) StuK40 L/43 gun (spring 1942) and later the improved 75mm (2.95in) StuK40 L/48 (autumn 1942). These later versions were known as the StuG III Ausf F, Ausf F/8 and Ausf. G.

Although the main armament was vital, it was recognized that the StuG III needed a more effective self-defence capability. Earlier variants had carried an MG34 in the fighting compartment, but lacked any firing mount, forcing the commander or loader to spray suppressive fire from an open hatch. This was highly inaccurate and left the machine gunner in an exposed position, but from December 1942, a square machine gun shield for the loader was fitted, which greatly improved both the accuracy of his fire and his chances of survival.

By mid-1943, *Das Reich's Sturmgeschütz Abteilung* (battalion) was equipped with the Ausf G variant. This was the final, and by far the most common, of the StuG series and was based on the hull of the Panzer III Ausf M. A widened upper superstructure was fitted and from May 1943, side hull skirts (*Schurzen*) were installed for added protection against Soviet anti-tank rifles. At the same time, 80mm (3.15in) thick plates were used for frontal armour instead of fitting 30mm (1.18in) appliqué armour to the basic 50mm (1.9in) frontal protection. The introduction of a rotating commander's cupola with periscopes allowed faster target acquisition and significantly improved the Ausf G's combat capability.

StuG Battalion: Equipment

UNIT	EQUIPMENT	STRENGTH
Signals Platoon	light command truck (HQ)	1
	Panzerbefehlswagen III command tanks	3
Pioneer Platoon:		
Platoon HQ	Motorcycle	1
	Kettenkrads	2
	Schwimmwagen	1
1 Reconnaissance Section	Motorcycle	2
	Kettenkrad	1
2 Reconnaissance Section	Motorcycle	2
	Kettenkrad	1
3 Reconnaissance Section	Motorcycle	2
	Kettenkrad	1
4 Reconnaissance Section	Motorcycle	2
	Kettenkrad	1
1 Engineer Section	SdKfz 251/7	1
	Maultier	1
2 Engineer Section	SdKfz 251/7	1
	Maultier	1
3 Engineer Section	SdKfz 251/7	1
	Maultier	1
AA Platoon		
Platoon HQ	Motorcycle	1
	Kübelwagen	1
1 AA Section	SdKfz 10/5	1
	Maultier ammunition carrier	1
2 AA Section	SdKfz 10/5	1
	Maultier ammunition carrier	1
3 AA Section	SdKfz 10/5	1
	Maultier ammunition carrier	1
1 StuG Company		
Company HQ	Motorcycle	3
	Kfz 15 radio car	1
	StuG III	2
1 Platoon	StuG III	3
2 Platoon	StuG III	3
3 Platoon	StuG III	3
2 StuG Company		
Company HQ	Motorcycle	3
	Kfz 15 radio car	1
	StuG III	2
1 Platoon	StuG III	3
2 Platoon	StuG III	3
3 Platoon	StuG III	3
3 StuG Company		
Company HQ	Motorcycle	3
	Kfz 15 radio car	1
	StuG III	2
1 Platoon	StuG III	3
2 Platoon	StuG III	3
3 Platoon	StuG III	3

StuG Battalion

Opel Blitz truck: 1
Motorcycle: 19
Kettenkrad: 6
Schwimmwagen: 1

SdKfz 251/7: 3
Maultier: 6
SdKfz 10/5: 3
Panzerbefehlswagen III: 3

Kfz 15 radio car: 3
StuG III Ausf G: 33
Kübelwagen: 1

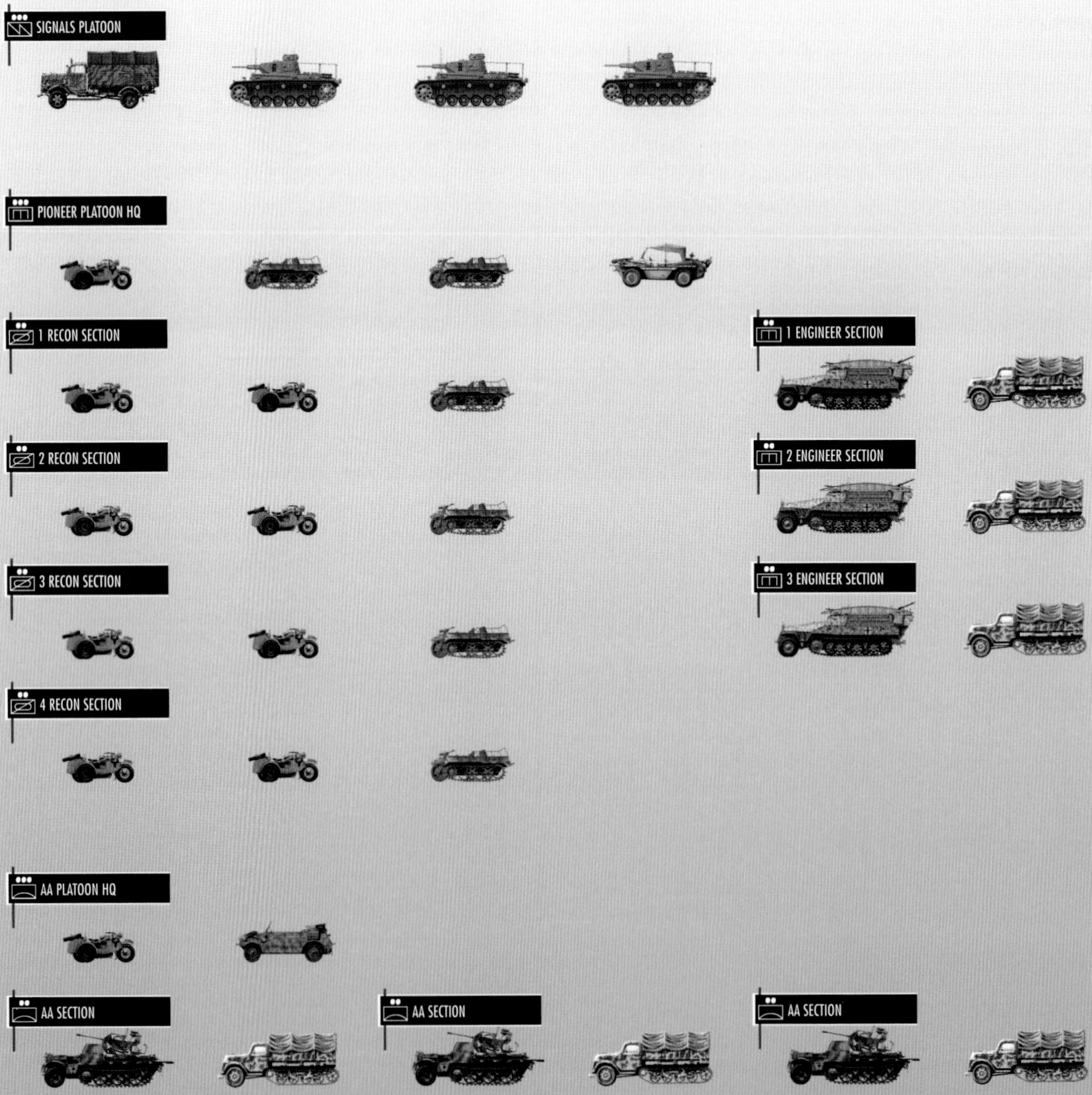

SIGNALS PLATOON

PIONEER PLATOON HQ

1 RECON SECTION

1 ENGINEER SECTION

2 RECON SECTION

2 ENGINEER SECTION

3 RECON SECTION

3 ENGINEER SECTION

4 RECON SECTION

AA PLATOON HQ

AA SECTION

AA SECTION

AA SECTION

By mid-1943, the StuG III had evolved from a pure infantry support AFV into a potent 'all-rounder', which was equally effective as a tank destroyer. Although its main armament lacked all-round traverse, its low silhouette, thick frontal armour and powerful gun made it one of the most formidable German AFVs.

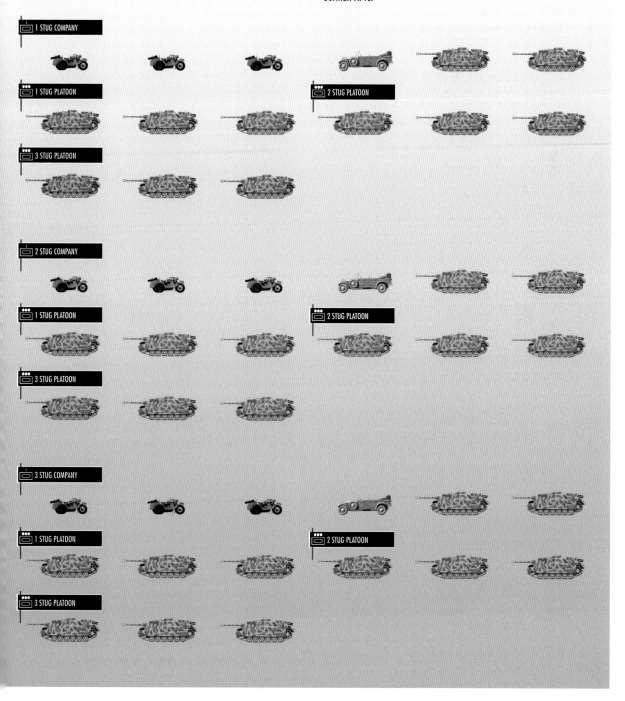

1 STUG COMPANY

1 STUG PLATOON

2 STUG PLATOON

3 STUG PLATOON

2 STUG COMPANY

1 STUG PLATOON

2 STUG PLATOON

3 STUG PLATOON

3 STUG COMPANY

1 STUG PLATOON

2 STUG PLATOON

3 STUG PLATOON

Medical Services

German military medical services were generally good and were far superior to those of the Red Army, if less comprehensive than their British and US counterparts.

The British and US armies made extensive use of the new medical practices and drugs of the era, including active immunization against tetanus, sulphonamide drugs and penicillin. In contrast, the Germans lacked antibiotics, with the result that many of their casualties suffered from severely infected wounds.

The efficiency of medical administration was also an important factor. This was particularly the case with the system for blood transfusion – in Britain, the well-organized Army Blood Transfusion Service was set up in 1938. This was in marked contrast with the German services, which routinely carried out blood transfusions from the start of the war but which had made inadequate arrangements for the collection and storage of blood in forward areas.

German medical services on the Eastern Front often worked under appalling conditions – the enormous distances and extremes of climate made it incredibly difficult to provide effective treatment for the large number of casualties sustained even on 'quiet' sectors of the line. In many cases the only hope for the seriously wounded was that they might be allocated a place on one of the over-worked casualty evacuation aircraft, usually the Junkers Ju 52S (*Sanitätsflugzeug* – ambulance aircraft).

SS *Das Reich* Medical Services

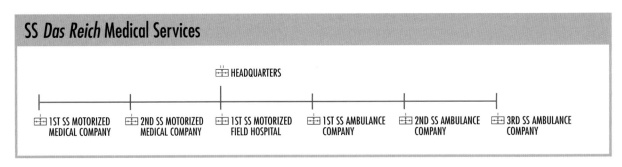

HEADQUARTERS

| 1ST SS MOTORIZED MEDICAL COMPANY | 2ND SS MOTORIZED MEDICAL COMPANY | 1ST SS MOTORIZED FIELD HOSPITAL | 1ST SS AMBULANCE COMPANY | 2ND SS AMBULANCE COMPANY | 3RD SS AMBULANCE COMPANY |

SdKfz 251/8

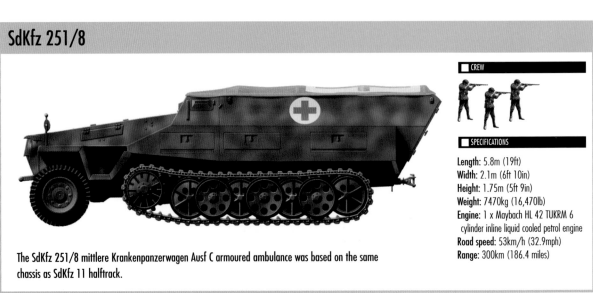

CREW

SPECIFICATIONS

Length: 5.8m (19ft)
Width: 2.1m (6ft 10in)
Height: 1.75m (5ft 9in)
Weight: 7470kg (16,470lb)
Engine: 1 x Maybach HL 42 TUKRM 6 cylinder inline liquid cooled petrol engine
Road speed: 53km/h (32.9mph)
Range: 300km (186.4 miles)

The SdKfz 251/8 mittlere Krankenpanzerwagen Ausf C armoured ambulance was based on the same chassis as SdKfz 11 halftrack.

UNIT MEDICAL SERVICES

These were controlled by the Medical Officer (*Abteilungsarzt*). Whenever possible, a doctor would move up with the forward troops in an armoured ambulance (usually an SdKfz 251/8), with medical orderlies in each platoon. If there were large numbers of casualties, the unit's supply vehicles might be commandeered to supplement its ambulances.

A US report on German army medical services was originally printed in *Tactical and Technical Trends* No 35, 7 October 1943 (see below and page 119). Whilst it refers to German army medical services, those of the *Waffen-SS* would be identical.

Evacuation of Casualties

UNIT	EQUIPMENT
Units under Divisional control	There are no medical units allotted to corps normally.
Battalion aid station (*Verwundetennetz*)	As close to the fighting line as possible. The station may or may not be under cover. Treatment is restricted to first aid. Occasional blood transfusions may be done. Evacuation is by stretcher bearer section of medical company.
Ambulance Station (*Wagenhalteplatz*)	This is established only if the ambulances cannot go forward to the battalion aid station.
Field Clearing Station (*Hauptverbandplatz*)	This is established by the medical company. It is intended for serious casualties requiring resuscitation, adjustment of dressings, splints etc, arrest of hemorrhage or blood transfusion before further evacuation. Emergency operations may be done here.
Lightly Wounded Collecting Station (*Leichtverwundetensammelplatz*)	For walking wounded and sitting or reclining cases requiring only minor treatment before evacuation. There is a skeleton staff from medical company. The remainder of the medical company is normally in reserve – available to assist in handling a sudden rush of cases or assist in bombed areas, etc.
Division Field Hospital (*Feldlazarett*)	Capacity 200 beds. It is intended for the reception and retention of casualties who require urgent operation or resuscitation and a few days rest before further evacuation. It has a surgical team and is fully equipped to handle any casualty.
Hospitals at Home or in Occupied Countries (*Reservelazaretten*)	
Casualty Collecting Station (*Krankensammelstelle*)	It is established by an Army ambulance unit at a railhead or other traffic centre, normally, and is for the retention of casualties awaiting evacuation. Only minor treatment is possible.
Army Field Hospital (*Kriegslazarett*)	For more serious casualties. Capacity 500 beds. Fully-equipped hospital with all specialist departments.
Army Field Hospital for lightly wounded cases (*Leichtkrankenkriegslazarett*)	Takes casualties who will be fit for duty in 3 or 4 weeks. Capacity 1000 beds; fully equipped. Normally located in back areas of Army zone and away from all large towns.
	Capacity and time of erecting or dismantling of various units
Army Field Hospital	Capacity 500 beds, 24 hours to set up or dismantle. Set up by Army medical detachments.
Army Field Hospital for slightly wounded cases	Capacity 1000 beds. 24 hours to set up or dismantle. Set up by Army field hospital detachments.
Field Hospital	Capacity 200 beds. 3 hours to set up or dismantle. Set up by Army medical detachments.
Casualty Collecting Station medical railhead	Capacity limited only by the available accommodation. 3 hours to set up or dismantle. Set up by motor transport ambulance company.
Field Dressing Station	Unlimited capacity. ½ to 1 hour to set up or dismantle. Set up by divisional medical company.
Slightly Wounded Collection Station	Unlimited capacity. A few minutes to set up or dismantle.
Motor Ambulance	Capacity of 4 lying and 10 sitting. Has four driving wheels and double differential for cross country performance.
Hospital Train	The 2- or 3-axled coach train with heating coach has 358 lying capacity, and 385 sitting capacity without heating coach. The 4-axled corridor coach train with or without heating coach has a 364 lying capacity. To set up or dismantle, 2 to 6 hours.
Hospital Train for Slightly Wounded	920 sitting capacity. To set up or dismantle, 1 to 2 hours.

Organization

DIVISION	UNIT	STRENGTH
Infantry Division	medical companies, motorized or partly motorized	2
	ambulance platoons, motorized	2
	field hospital, motorized or partly motorized	1
Mountain Division	medical companies, partly motorized	2
	ambulance platoons, motorized	2
	field hospital, partly motorized	1
Motorized Division	medical companies, motorized	2
	ambulance platoons, motorized	3
	field hospital, motorized	1
Armoured Division	medical companies, motorized and armoured	2
	ambulance platoons, motorized and armoured	3
Airborne Division	paratroop medical companies	3
	field hospital (airlanding)	1

Equipment of the Combat Unit

UNIT	EQUIPMENT
Box No. 1	This is the Battlebox (*Gefechtskasten*) and is marked with two white strokes in the form of a cross. The contents are miscellaneous but include much dressing material, anti-tetanus serum, medicaments and two 6-in atomizers containing ethyl chloride.
Box No. 2	Medicaments of all types (*Arzneimittelkasten*).
Box No. 3	Dressing Equipment (*Verbandmittel*).
Box No. 4	Supplementary box. This is like No. 1, but on a smaller scale.
Box No. 5	Contains 280 flasks of tetanus anti-toxin, 3000 units per cc.

Two medical haversacks containing medicaments and dressings rather on the scale of the MO's haversack. Two empty rucksacks are also included, with blocks of labels for wounded and for sick.

One set of equipment for fractures, including cardboard splints, metal wire splints and aluminum splints.

One unit medical outfit – like an MO's haversack.

One set of oxygen apparatus – the flask contains 275 quarts of oxygen.

Four stretchers.

Twelve woolen blankets.

One filter apparatus.

Anti-vesicant, and gas protection caps for those with head wounds.

Equipment of the Individual

UNIT	EQUIPMENT
The Combatant	One large and one small packet first field dressing wrapped separately in black, rubberized fabric. 1 box anti-vesicant tablets
The Stretcher Bearer	In the combatant units – 2 stretcher-bearer's haversacks containing: 1 pair of scissors 1 pair of dissecting forceps 6 packets dressing 3 triangular bandages 3 strips of gauze 16ft x 2 ¾ in (5m x 7cm) 6 squares of gauze 1 roll of adhesive tape 1 waterproof bandage 18 x 20in (45 x 50cm) 1 tourniquet 20 safety pins
In the medical units	In this case, the stretcher bearers are not all equipped with the haversack as above, but each group of four has a haversack (*Verbandtasche*) containing: 1 pair of cloth-cutting scissors (*Kleiderschere*) 1 tourniquet 12 strips of gauze 10 squares of absorbent cotton 6 triangular bandages 2 *Brandbinden* (absorbent gauze treated with bismuth for burns) 1 waterproof bandage 36 x 40in (90 x 100cm) 4 rolls rubberized adhesive tape 35 safety pins 4 small splints with cradle 2 slings, 12 x 5in (30 x 12.5cm)
	Each stretcher bearer also carries a mug and a bottle with about a pint of cordial.
The Medical Service NCOs and Medical Orderlies	Each carries a bottle of cordial, a case of dressing material and two medical haversacks, the first containing medicines such as salicylic acid (2%), formaldehyde, ticture of iodine, cardiazol and opium; the second containing much the same dressing material as the stretcher bearers' haversack on a smaller scale.
Officers of the Medical Services	Each carries the officer's haversack, the contents of which are extensive, including a number of surgical instruments such as probes, lancets, ligature forceps, cannulae, vaccinostyles and the like; a certain quantity of dressings; a tin plate case for tablets containing among others, pyramidone, veronal, acetylsalicylic acid, codein phosphate, tannalbin, opium, cocain chloride, atropinemannite and calomel; and also a box containing ampoules of, for example, caffein sodium salicylate, superarenine chloride and morphin hydrochloride.

Phänomen Granit 1500A Ambulance

CREW

SPECIFICATIONS
Weight: 3.75 tonnes (3.69 tons)
Length: 5.5m (18ft)
Width: 1.98m (6ft 6in)
Height: 2.08m (6ft 10in)
Engine: Air-cooled 4-cylinder diesel
Speed: n/k
Range: 430km (375 miles)

The Granit was one of the commonest German military ambulances, equipping the majority of medical units.

Die Logistik (Logistics)

A rough translation of a German military saying goes, 'Supply isn't everything, but without supply everything else amounts to nothing.' Whilst hardly the most memorable catchphrase, it sums up the fact that a powerful armoured force has no more value than scrap metal unless it has essential supplies (primarily fuel and ammunition).

At company level, supply and support elements were designated *Gefechtstrossen* (combat trains). In operational areas, two such trains were generally formed, designated *Gefechtstross* I and *Gefechtstross* II. *Gefechtstross* I was commanded by an experienced NCO and contained the only the essential support elements to maintain the unit's combat capability, usually:

- Refuelling vehicles
- Ammunition carriers
- *Bergefahrzeuge* (recovery vehicles)
- Vehicles for technical and weapon specialists and their equipment
- Replacement crews
- *Fieldkuchen* (field kitchen)

On the march, each company's *Gefechtstross* I moved up with the leading combat elements, but on deployment it merged with the combat trains of the battalion's other companies in the battalion's rear area. It was then held in readiness to be brought forward to resupply the unit's combat elements before they went into action. All remaining logistic elements of each company formed *Gefechtstross* II vehicles under the command of an *SS-Hauptscharführer* (Company Sergeant Major).

One of the worst logistic problems lay in the inability of German war industries to produce sufficient numbers of trucks suitable for military use in the harsh conditions of the Eastern Front. Whilst favoured units such as *Das Reich* generally received reasonable numbers of good vehicles such as the Opel Blitz and Maultier, other motorized formations had to make do with an amazing variety of captured and requisitioned vehicles, many of which suffered from inherent mechanical unreliability and shortages of spare parts.

Although at unit level, *Waffen-SS* units seem to have operated much like their army counterparts, their supply chain was unique. The US War Department Technical Manual, *TM-E 30-451: Handbook on German Military Forces*, published in March 1945, summarized its organization as follows:

SS *Das Reich* Supply Service

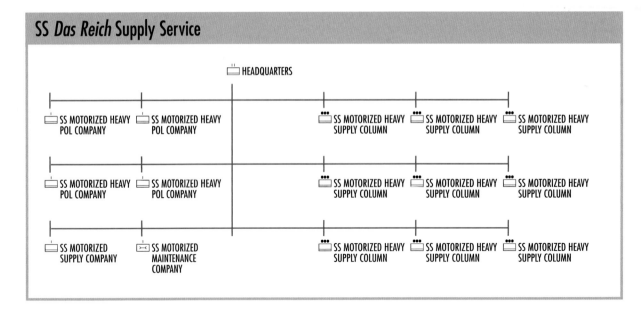

(1) General. Units of the Waffen-SS operating under the tactical control of the Army utilize the regular Army supply channels for supplies of rations, fuel, heavy equipment and ammunition. In addition, however, the SS maintains its own system of supply distinct from that of the Armed Forces and not subject to control or supervision by the latter. For this purpose a large network of depots and stores has been built up in Germany and in occupied territory.

(2) Control. Operationally these depots and stores come under the control of the SS Main Operational Department, which is responsible for the equipment and supply of SS units and establishments when not under the tactical control of the Army. The SS Main Economic Administrative Department, on the other hand, is responsible for the detailed administration of these depots, or for the general supervision of administration where there is decentralization of its authority, e.g. to the economic official (*SS-Wirtschafter*) with a Higher SS and police commander in occupied territory. The actual responsibility for supply is divided between the SS Main Operational Department, which is responsible for initial equipment and the supply and maintenance of arms, ammunition, technical equipment, and transport vehicles, and the SS Main Economic Administrative Department, which is responsible for rations, clothing, personal equipment, coal, wood, and fodder.

For certain types of heavy equipment which are obtainable only from Army depots, agreement is reached between the OKH and the SS Main Operational Department, which becomes responsible for the general supervision of stocks, maintenance, and repair once such equipment has been handed over to an SS unit.

Although the SS and Police supply and administration system in wartime operates primarily for the Waffen-SS and SS police units, its organization and installations are also at the disposal of the General SS and the SS Death's-Head Formations.

(3) Regional organization of supply. The SS Main Economic Administrative Department controls all regional supply depots. There is a marked tendency for SS depots and administrative services to be grouped around concentration camps, notably Dachau and Oranienburg. This arrangement centralized administrative matters, as the concentration camps come under the control of the SS Main Economic Administrative Department, and the inmates of such camps provide a cheap source of labor.

At each SS district headquarters in Germany proper there is an administrative office (*Verwaltungsamt*) which controls and supervises all supply depots and installations within its area. Similarly at the SS sub-district headquarters there is an administrative branch (*Verwaltungsabteilung*) with the same functions. At the headquarters of an SS garrison command (*SS-Standortbereich*) there is an SS garrison administration headquarters (*SS-Standortverwaltung*) dealing with supply and finance in its area and directly subordinate to the respective SS district and sub-district.

In occupied territory, there is an economic section controlled by an official (*SS-Wirtschafter*) on the staff of an HSSPf. He is responsible for the administration of all depots and supplies in his region. Where field units of the Waffen-SS are likely to operate in a particular area for a considerable period, special supply bases (*Stutzpunkte*) are usually established at convenient points. These are small and temporary in character.

(4) **Channels of supply.** All Waffen-SS units requisition their supplies from the SS Main Operational Department, which either makes the issue itself or instructs the SS Main Economic Administrative Department to do so. The latter then either dispatches the material direct to the unit from one of the central depots or from the factory, or arranges for it to be made available to the unit at the nearest convenient sub-depot.

The main stocks of supply are held in central SS depots. These are of two kinds:

Main supply depots (*SS-Hauptwirtschaftslager* – HWL), containing miscellaneous types of supplies.

Special depots, including SS ordnance depots, motor transport supply depots and parks, signal equipment depots, medical equipment depots, and clothing depots.

From these central depots, outlying sub-depots are supplied. These may be either SS supply depots (*SS-Nachschublager*), mainly found near the borders of Germany and in occupied territory, or SS troop supply depots (*SS-Truppenwirtschaftslager* – TWL), which hold stocks of clothing, light equipment, fuel, and other goods.

At the time when the supply lines on the Eastern Front were too far extended, SS supply service headquarters (*SS-Nachschubkommandanturen*) were established. Each of these was in itself an important group of depots and administrative offices. Although subordinate for administrative purposes to the SS economic official with the local HSSPf, it was the primary link between the SS

Maultier

CREW

SPECIFICATIONS

Weight: 7.1 tonnes (6.98 tons)
Length: 6.02m (19ft 9in)
Width: 2.265m (7ft 5in)
Height: 2.5m (8ft 2in)
Engine: 54.8kW (73.5hp) Opel 6-cylinder petrol
Speed: 38km/h (23.6mph)
Range: n/k

Maultier (Mule) was the generic name of a series of halftracks based on Opel, Mercedes-Benz and Ford trucks. During 1941/42 it became apparent that conventional wheeled transport vehicles were unsuitable for the deep mud of the *rasputitsa* that marked the beginning and end of the Russian winter. Only halftracks and fully-tracked vehicles were able to operate in these conditions, but there were virtually none to spare for supply duties. One of *Das Reich's* field workshops experimentally removed the rear suspension and wheels of a standard 3-ton truck and replaced them with the running gear of a captured Soviet light tank. This improvisation was so successful that a programme was set up to convert standard trucks to halftrack configuration and a total of approximately 6000 were completed between 1942 and 1945.

main departments and main depots in Germany and the SS units and sub-depots in its own area. It served both as a distribution centre and a supply base, and in its depots were held arms, ammunition, transport equipment, captured material, clothing, fuel, coal, wood and building materials. It was also empowered to make contracts with or purchases from private firms in its area.

(5) **Veterinary supply service.** The Waffen-SS maintains its own channel of supply for its cavalry and non-motorized units. Horses for the Waffen-SS are procured through SS remount depots (*SS-Remonteämter*), which were mainly found until recently in occupied territory. These depots forward the horses to the SS riding and driving schools (*SS-Reit-und-Fahrschulen*), from where they either go to a unit direct or to an SS base veterinary depot (*SS-Heimatpferdepark*), which in turn forwards them to an SS veterinary depot (*SS-Pferdepark*) in a forward area, usually attached to an SS corps. These corps will then make distribution among their divisions which have veterinary companies. Wounded horses, after

treatment in the field, go to an SS veterinary hospital (*SS-Pferdelazarett*) in a forward area and then to an SS base veterinary hospital (*SS-Heimatpferdelazarett*) in Germany. Veterinary equipment for Waffen-SS units can be obtained from the SS Central Veterinary Park (*SS-Hauptveterinärpark*) by way of one of the veterinary parks in forward areas.

(6) **Movement of supplies.** The transportation of SS supplies is coordinated by the transportation officer (*Transportoffizier* – TO) in the SS Main Operational Department. He maintains liaison both with other SS main departments where necessary and also with the German railway authorities and the transportation authorities of the German Army. Subordinate to him are a number of regional transportation officers, found mainly in those districts close to the German border. Other transportation officers are stationed at principal railway stations in Germany and in occupied territory. At railway junctions particularly important for SS movement, SS reloading stations (*SS-Umschlagstellen*) are established.

Opel Blitz 3000 Type S

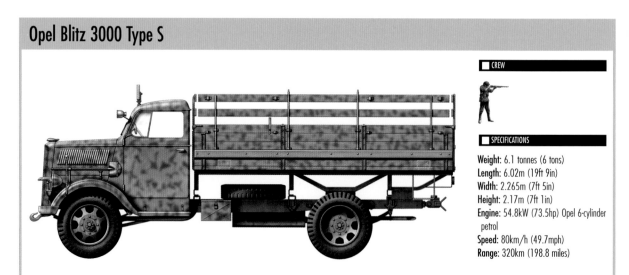

CREW

SPECIFICATIONS

Weight: 6.1 tonnes (6 tons)
Length: 6.02m (19ft 9in)
Width: 2.265m (7ft 5in)
Height: 2.17m (7ft 1in)
Engine: 54.8kW (73.5hp) Opel 6-cylinder petrol
Speed: 80km/h (49.7mph)
Range: 320km (198.8 miles)

By the late 1930s, the rapid expansion of the *Wehrmacht* had led to an inventory of over 100 different types of vehicle. Even under peacetime conditions, it was proving almost impossible to provide adequate logistic support for such a varied collection of vehicles, and a rationalization programme began with the aim of reducing the number of types in service to no more than 30. The Opel Blitz was one of the most successful models to be adopted under this programme, with over 70,000 produced between 1937 and 1944. Besides the standard truck bodies, the chassis was used for a wide variety of specialist roles, including ambulances and fuel tankers.

Instandsetzungsgruppen (Repair and Maintenance Sections)

These were amongst the most important non-combatant elements of any German armoured force. Their efficiency in recovering and repairing knocked-out AFVs played a key role in maintaining a unit's combat strength.

Their recovery equipment was as good as any fielded by Allied armies until the middle of the war, when the introduction of heavier AFVs such as the Tiger and Panther posed considerable problems.

The standard SdKfz 9 'Famo' recovery halftrack tractor had a designed towing capacity of 28 tonnes (27.55 tons). For moving disabled tanks over longer distances, it towed the SdAnh 116 low-loader trailer, which could carry AFVs of up to 24 tonnes (23.6 tons).

This was adequate for AFVs such as the Panzer IV, but it was found that three tractors were usually needed to tow a single immobilized Tiger I. Initially, another Tiger would often attempt the job, which frequently resulted in the engine of the towing vehicle overheating and breaking down or even catching fire. This led to an official ban on using Tigers as towing vehicles, which was frequently ignored in emergencies.

WORKSHOP COMPANY

Things became much easier once the disabled AFV arrived at the workshop company, which had the facilities to carry out all but the most extensive repairs. These included gantry cranes capable of lifting even the heaviest components such as turrets or engines, supplemented by the SdKfz 9/1

which mounted a 6-tonne (5.9-ton) capacity crane and the petrol-electric 10-tonne (9.8-ton) crane on the later SdKfz 9/2.

Such comprehensive facilities were matched by the expertise of the workshop company personnel. In addition to the highly skilled mechanics, there were frequently civilian experts seconded from the tank factories who were able to assist with particularly complex repairs.

The workshops' most serious problems were frequent shortages of spare parts – these were largely due to Hitler's obsession with increasing the numbers of new tanks built, whilst refusing to give sufficient priority to the manufacture of spares. Guderian's repeated advice that an adequate supply of spares would increase the *Panzerwaffe*'s combat strength far more quickly and cheaply than building new AFVs went unheeded, largely because Hitler would not accept the 20 per cent reduction in AFV production that this would entail.

RECOVERY AND REPAIR

Despite these many obvious difficulties, the *Instandsetzungsgruppen* achieved remarkable results throughout the Kursk offensive, aided by the fact that German forces were advancing, which made it relatively easy to recover vehicles for repair. The long

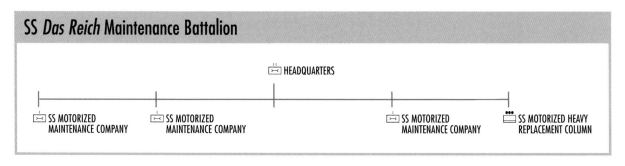

SS *Das Reich* Maintenance Battalion

HEADQUARTERS

SS MOTORIZED MAINTENANCE COMPANY

SS MOTORIZED MAINTENANCE COMPANY

SS MOTORIZED MAINTENANCE COMPANY

SS MOTORIZED HEAVY REPLACEMENT COLUMN

series of retreats that began after Operation 'Citadel' saw a dramatic rise in AFV losses because disabled vehicles could not be retrieved and all too frequently had to be blown up to prevent their capture.

The table information (below and opposite) is based on a US Army report, published in *Tactical and Technical Trends* No 10 of 22 October 1942. Although the reports on panzer maintenance and recovery refer to German army *Instandsetzungsgruppen* and associated units, their *Waffen-SS* equivalents were organized along very similar lines.

Company Repair Section Equipment

EQUIPMENT	STRENGTH
NCO (tank mechanic), section leader	1
NCOs, tank mechanics	3
privates, tank mechanics	13
privates, tank radio electricians	2
private, armourer's assistant	1
privates, chauffeurs	4
small repair car (Kfz 2/40)	1
medium cross-country repair truck, for spare parts and tools	1
halftrack vehicles (SdKfz 10) for personnel, capable of towing 1 ton	2
Motorcycle with sidecars	3

Battalion Repair Section Equipment

EQUIPMENT	STRENGTH
NCO (tank mechanic), section leader	1
privates, tank mechanics (for a tank regimental HQ) or	3
privates, tank mechanics (for a tank battalion HQ)	5
private, motorcyclist, tank radio electrician	1
private, chauffeur, tank radio electrician	1
chauffeur	1
small repair car (Kfz. 2/40)	1
medium cross-country repair truck, for spare parts and tools	1
Motorcycle with sidecar	1

SdKfz 9

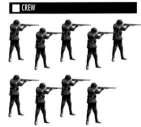

CREW

SPECIFICATIONS

Weight: 18 tonnes (17.72 tons)
Length: 8.32m (27ft 3.5in)
Width: 2.6m (8ft 6in)
Height: 2.85m (9ft 4in)
Engine: 186.42kW (250hp) Maybach HL 108 TUKRM 12-cylinder petrol
Speed: 50km/h (31mph)
Range: 260km (161.55 miles)

Development of this heavy halftrack began in the early 1930s and the first prototype underwent trials in 1936. The vehicle entered service in 1939 and was initially used as a heavy artillery tractor and recovery vehicle. Its towing capacity of 28 tonnes (27.56 tons) was adequate for all German AFVs in service until 1942, when the Tiger I became operational. It was soon found that up to three SdKfz 9 halftracks were needed to recover and tow a disabled Tiger I in difficult terrain and the problem was not fully solved until the Bergpanther ARV came into service in 1944. A number of vehicles were converted to carry a 6000kg (13,228lb) capacity crane (SdKfz 9/1). These proved to be so useful that a further batch were more extensively modified to mount a 10,000kg (22,046lb) capacity crane (SdKfz 9/2).

Workshop Company

UNIT	EQUIPMENT	MEN
Headquarters Platoon	1 cross-country truck (Kfz 1)	1 chauffeur, 1 company commander (engineer), 1 officer for special duties (engineer), 1 clerk (draftsman); (one of the two officers may be other than an engineer officer)
	1 motorcycle	1 motorcyclist (orderly)
	1 medium truck	1 chauffeur, 2 men for salvaging spare parts (M)
	1 light personnel car	1 chauffeur, 1 official (K-motor transport), 1 NCO for spare parts, 1 clerk (asst. chauffeur)
	1 motorcycle with sidecar	1 motorcyclist (orderly), 1 foreman for motor transport equipment (Maybach Specialist)
1st and 2nd Platoons	1 motor bus (*Kraftomnibus*)	1 chauffeur, 4 NCOs for workshop service
		1 tank electrician and mechanic, 1 tank electric welder, 1 saddler, 1 tinsmith, 1 carpenter, 1 painter, 7 tank motor mechanics, 3 tank transmission mechanics, 1 automobile mechanic, 1 clerk
	5 medium trucks, for spare parts and assemblies	(each) 1 chauffeur, 1 tank transmission mechanic (asst. chauffeur), 1 automobile mechanic
	1 medium truck for spare parts and assemblies	1 chauffeur, 1 NCO in charge of spare parts, 1 depot chief (M)
	1 truck with special workshop and trailer for arc-welding apparatus	1 chauffeur, 1 NCO for workshop service (*vorhandwk*) 1 tank electric welder (asst. chauffeur)
	1 heavy truck, tools and equipment	1 chauffeur, 1 tank motor mechanic, 1 blacksmith
	1 workshop truck (Kfz19), with trailer for heavy machine apparatus, Set A	1 chauffeur, 1 foreman (leader), 1 turner
3rd Platoon (Recovery Platoon)	1 light cross-country automobile (Kfz 1)	1 chauffeur, 1 officer (platoon leader), NCO (Panzer-Wart, tank mechanic)
	1 medium cross-country truck (Kfz 100) for towing apparatus, with rotating crane (3 tons)	1 chauffeur, 1 asst. chauffeur (automobile mechanic)
	1 medium half-track prime mover (8 tons)	1 chauffeur, 1 assistant chauffeur (automobile mechanic)
	2 medium half-track prime movers (8 tons) with underslung trailers (10 tons)	(each) 1 chauffeur, 1 asst. chauffeur (mechanic), and (for one only of these trucks) 1 NCO (tank mechanic)
	2 vehicles (with apparatus) (6 tons, SdKfz 41)	(each) 1 chauffeur, 1 assistant chauffeur (automobile mechanic)
	5 heavy half-track prime movers (18 tons), with underslung trailers (20 tons)	(each) 1 chauffeur, 1 assistant chauffeur (automobile mechanic), 1 steerer for trailer; one prime mover has in addition, an NCO (tank mechanic)
	2 motorcycles with sidecars	(each) 1 chauffeur (tank mechanic), 1 NCO (tank mechanic) (One of the NCOs is second in command.)
Armoury Section	1 medium cross-country automobile (Kfz 15 m.G.)	1 chauffeur, 2 armourers (one is section leader), 1 armourer's helper
	1 motorcycle with sidecar	1 NCO armourer (0), 1 helper
	3 vehicles (not described), for armourer's tools	One with 1 chauffeur, 1 NCO, armourer, 1 tank electrician and mechanic (asst. chauffeur); One with 1 chauffeur, 1 tank electrician (asst. chauffeur), 1 armourer's helper; One with 1 chauffeur, 2 armourer's helpers (one is asst. chauffeur)
	1 light cross-country car for supply of tools	1 chauffeur, 1 armourer's helper
Workshops for Communications Equipment	1 battery-charging truck (Kfz. 42)	1 chauffeur, 1 NCO mechanic (leader), 1 mechanic
	1 communications workshop truck (Kfz. 42)	1 chauffeur, 1 mechanic (asst. chauffeur)
	1 light cross-country truck	1 chauffeur, 1 mechanic (asst. chauffeur)
Company Supply	1 medium truck for rations and baggage	1 chauffeur, 1 NCO in charge of equipment (leader)
	1 motorcycle with sidecar	1 supply sergeant (K), 1 clerk (asst. motorcyclist)
	1 anti-aircraft truck (Kfz. 4)	1 chauffeur, 1 NCO (in charge), 1 machine-gunner
	2 medium trucks for fuel	One, with 1 chauffeur and 1 tailor (asst. chauffeur); One, with 1 chauffeur and 1 shoemaker (asst. chauffeur)
	2 medium trucks for large field-kitchen stoves	One, with 1 chauffeur, 1 NCO in charge of rations (asst. chauffeur), 1 cook, 1 asst. cook; One, with 1 chauffeur, 1 NCO (accountant), 1 NCO (cook), 1 asst. cook (asst. chauffeur)
	1 light automobile	1 chauffeur (clerk), 1 master sergeant, 1 medical officer

Chapter 3
The *Luftwaffe* at Kursk

The *Luftwaffe* had been an essential element in the devastating
German victories of 1939 and 1940. In fact, it was only in the air that Germany had a
clear numerical and qualitative superiority during the Battle of France in May and
June 1940 – in terms of manpower, tanks and artillery, it was outnumbered
by the Allied forces.

Amazingly, despite the *Luftwaffe*'s key role in the success of what
the press dubbed *Blitzkrieg*, German aircraft production dismally failed to keep pace
with the demands of a steadily widening war. In 1940 the Soviet Union, still reeling
from the effects of Stalin's purges, produced 11,000 aircraft whilst Germany managed
barely 10,000. The following year, despite the massive disruption caused by the
German invasion, Soviet factories completed 16,000 warplanes compared
to Germany's 12,000 and pulled further and further ahead for
the remainder of the war.

As if this was not serious enough, Göring's incompetence and inability to appreciate
the importance of rapidly developing military aviation technology condemned the
Luftwaffe to reliance on increasing outdated aircraft whilst the Red Air Force fielded
completely new and more formidable warplanes.

OPPOSITE: A remarkable shot from the nose of a Heinkel He III, taken during the Kursk campaign. It was in the sky over the Kursk battlefield that the *Luftwaffe* finally lost its mastery on the Eastern Front.

Early Successes

In the opening stages of Operation *Barbarossa*, the *Luftwaffe* seemed as just as invincible as it was portrayed by Goebbels' propaganda. On the first day of the German offensive, 66 major airfields in western Russia were subjected to a series of devastating *Luftwaffe* attacks – 800 Soviet aircraft were destroyed on the ground whilst another 400 or so were shot down.

For much of the rest of the year, the *Luftwaffe* enjoyed complete air superiority over the entire Eastern Front. Its highly-trained fighter pilots routinely massacred the remnants of the Red Air Force, with many running up huge personal scores. German bomber and ground attack units were able to raid at will, with little opposition except for fierce, but generally inaccurate, AA fire. Initially the most demanding task was prioritizing targets – reconnaissance flights constantly identified so many Red Army concentrations that it was physically impossible to attack them all. Many had to be ignored so that those which posed the greatest

Although the Red Air Force's machines had improved in quality and its pilots in experience, they still endured heavy casualties in men and aircraft. Here a Yakovlev Yak 1M fighter plunges to the ground in flames.

threat were targeted. (The situation was worsened by a temporary shortage of bombs – a worrying foretaste of the production and supply problems that were to dog the *Luftwaffe* throughout the remainder of the war.)

The plethora of targets and ever-increasing demands from the army for close air support ruled out anything more than a token *Luftwaffe* effort against industrial and strategic targets. As a result, much of Soviet war industry in areas threatened by the German advance was dismantled and evacuated without suffering any significant damage from air attacks.

The *Luftwaffe*'s constant presence above the Eastern Front's battlefields in 1941 appeared to be a symbol of German technical superiority, but was in fact a symptom of the army's weakness. Only a

small proportion of German ground forces were mechanized and their artillery and anti-tank weapons were simply insufficient to destroy even the ill-led and largely demoralized Red Army of 1941. *Luftwaffe* close air support was essential to allow the panzer formations to achieve their breakthroughs and to seal off and destroy the huge pockets of Soviet troops trapped by their advance. Stukas and medium bombers were constantly in demand to act as 'flying artillery', whilst the over-worked transport squadrons were frequently diverted from other essential tasks to fly in emergency supplies of fuel and ammunition to allow the panzer spearheads to remain operational.

As the campaign in Russia continued, the *Luftwaffe* struggled to operate effectively. Aircrew attrition steadily worsened, forcing the operational deployment of training units, which marked the first stage of a progressive decline in aircrew quality as training began to be disrupted. This qualitative decline was worsened as training courses were gradually shortened to provide aircrews more rapidly. A lack of maintenance capability, long and inefficient supply lines, bad weather and primitive operating conditions rapidly reduced the *Luftwaffe*'s operational capabilities. By the end of December 1941, only 54 per cent of available

single-engine fighters and 32 per cent of bombers remained operational.

In early 1942, the *Luftwaffe* was heavily committed to supporting German forces trapped in the Demyansk and Kholm pockets, which were able to survive only thanks to major air resupply operations. During the four months of the two airlifts, more than 65,000 tons (63,970 tonnes) of supplies and 31,000 reinforcements were flown in whilst 36,000 wounded were evacuated on the return flights. However the cost was high, with the *Luftwaffe* losing 265 invaluable transport aircraft.

In the summer of 1942, the *Luftwaffe*'s aircraft provided decisive support to von Manstein's forces in the Crimea. The whole of *Luftflotte* 4 was deployed in support of the assault on Sevastopol, flying 23,000 sorties over a three-week period, during which it pounded the Soviet defences with 20,529 tons (20,200 tonnes) of bombs. (In comparison, the *Luftwaffe* dropped a total of 21,860 tons/ 21,510 tonnes of bombs on Britain throughout the entire 'Blitz' of 1940/41.) By the beginning of July, resistance in the Crimea had collapsed and the majority of aircraft were moved north to support the main German offensives against Stalingrad and the Caucasus. On 23 August, *Fliegerkorps* VIII began operations against Stalingrad. In the first week of the

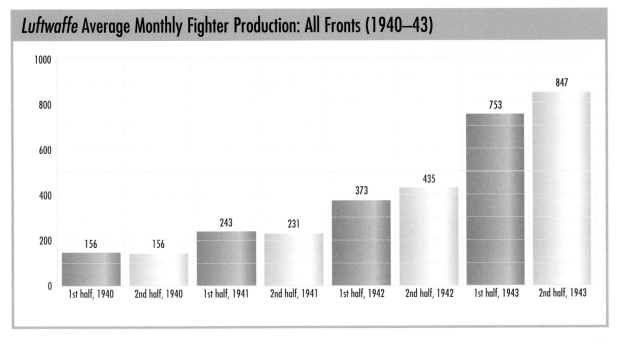

Luftwaffe Average Monthly Fighter Production: All Fronts (1940–43)

1st half, 1940	156
2nd half, 1940	156
1st half, 1941	243
2nd half, 1941	231
1st half, 1942	373
2nd half, 1942	435
1st half, 1943	753
2nd half, 1943	847

assault on the city, its aircraft flew 1600 sorties dropping more than 984 tonnes (1000 tons) of bombs.

The winter of 1942/43 saw the *Luftwaffe* on the Eastern Front struggling to maintain the Stalingrad airlift. Appalling weather conditions and aircraft flying from temporary airfields, at the limit of their range,

reduced the effectiveness of the operation. Moreover, the *Luftwaffe* did not have enough suitable aircraft to support Sixth Army trapped inside the pocket, which required a daily minimum of 787 tonnes (800 tons) of supplies – the average daily delivery totalled just 92.5 tonnes (94 tons) at a cost of 495 aircraft.

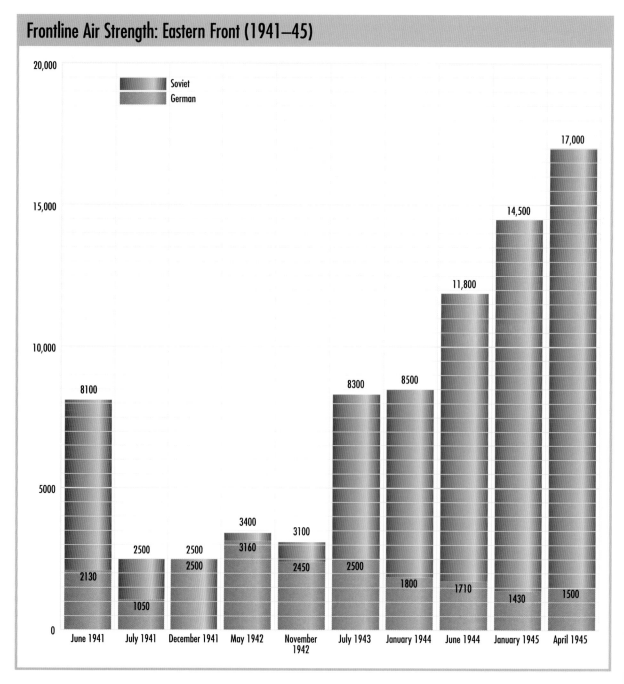

Frontline Air Strength: Eastern Front (1941–45)

Soviet
German

Operation 'Citadel'

In the final weeks of preparation in late June, the *Luftwaffe* began concentrating on attacks against the Red Air Force and interdicting the rail lines in the immediate area of Kursk.

Overall, the air interdiction campaign did not have a significant result on Soviet capability in the forthcoming battle. This was largely because, in spite of the massive effort mounted by the *Luftwaffe,* insufficient forces were employed to effectively damage the Soviet rail system.

In analyzing the results of the interdiction campaign, *Generalleutnant* Hermann Plocher wrote:

'Because the *Luftflotte* 6 (and *Luftflotte* 4) lacked sufficient force … and because [the] operational area was so vast, it was unable to seal off the area or to annihilate the enemy concentrations. Very broad missions were assigned to the two *Luftflotten.* They were to use long-range aircraft to interdict Russian communications and provide close air support to aid the [attacking] Army Groups in making narrow front penetrations. … *Luftflotte* 6 in the north was initially to attack Soviet airfields and artillery concentrations throughout the salient prior to the opening of the ground offensive. As the battle progressed, attacks were to be shifted to support XXXXVII Panzer Corps, which was Ninth Army's breakthrough force and Second Panzer Army, which was defending in the remainder of the Orel salient. An added mission for the *Luftflotte* 6 was to "patrol the entire assault area". *Luftflotte* 4 in the south was initially tasked with gaining air superiority over the line of advance. Emphasis was then to be shifted to providing general support for all elements of Fourth Panzer Army and Army Detachment Kempf. Additionally, close contact was to be maintained with II SS Panzer Corps, the southern … breakthrough force.'

Luftflotten 4 and 6 controlled more than 1800 of the 2500 German aircraft of all types on the entire Eastern Front. *Luftwaffe* units were released from duties on other portions of the front and replaced with less capable Rumanian and Hungarian units, which were given defensive missions. Throughout the operation the *Luftwaffe* suffered from a shortage of fully trained personnel, which placed tremendous demands on experienced aircrews who were rapidly exhausted by flying four, five or six sorties per day.

Fuel shortages were to plague the *Luftwaffe* even before the offensive began. In June 1943, *Luftflotte* 6 reported that total monthly consumption of B4 (89-octane aviation fuel used by bombers and most other aircraft) amounted to 8634 tons (8498 tonnes), whilst deliveries totalled only 5722 tons (5632 tonnes). During the same month, Fw 190s had used 1079 tons (1062 tonnes) of C3 (94-octane aviation fuel), but only 441 tons (434 tonnes) had been received. The huge quantities of fuel required were most efficiently transported by rail, but the time taken to move fuel trains from the Reich to forward airfields was at least two days and more often four to eight days. Such delays were largely due to partisan attacks against rail lines. These attacks peaked in Army Group Centre's rear areas during June, when partisans carried out 841 attacks, damaging 298 locomotives, 1222 rail wagons and 44 bridges. On any given day, partisan attacks in this sector damaged the tracks at an average of 84 separate points.

These factors combined to place extraordinary constraints on the *Luftwaffe,* as fuel came to dictate the pace of operations, forcing the introduction of 'fuel tactics'. Requests for air support had to be ruthlessly prioritized and granted in only the most critical situations, with *Luftwaffe* planning staff vetting every sortie to determine whether it was truly worth the fuel expenditure.

SOVIET DEFENCES

The Red Air Force opened the air campaign on 6 May, when waves of 300–400 aircraft from several air armies attacked *Luftwaffe* airfields around the Kursk salient over a three-day period. (At the time, *Stavka* anticipated that Operation 'Citadel' would be launched no later than 12 May and ordered the raids in the hope of crippling the *Luftwaffe* before the offensive.) Although the initial attacks achieved a measure of surprise and damaged or destroyed 22 *Luftwaffe* aircraft on the ground, the later raids were mauled by thoroughly alerted German defences, which shot down a total of 101 Soviet aircraft.

Fliegerkorps VIII (*Generalmajor* Hans Seidemann)

UNIT	AIRFIELD	COMMANDER	EQUIPMENT	STRENGTH
Stab/KG 3		OTL Walter Lehwess-Litzmann	Ju 88	1
I./KG 3	Stalino and Poltava	Major Joachim Jodicke	Ju 88	34
II./KG 3	Poltava	Major Jurgen de Lalande	Ju 88	37
Stab/KG 27		OTL Hans-Henning Freiherr von Beust	He 111	2
I./KG 27		Hptm. Joachim Petzold	He 111	21
II./KG 27		Major Karl-August Peterson	He 111	34
III./KG 27	Kharkov–Voychenko	Hptm. Karl Mayer	He 111	34
14.(Eis.)/KG 27			He 111	16
Stab/KG 55	Kharkov–Rogan and Stalino	OTL Ernst Kuhl	He 111	4
II./KG 55	Kharkov–Rogan and Stalino	Major Heinz Hofer	He 111	32
III./KG 55	Kharkov–Rogan and Stalino	Major Willi Antrup	He 111	46
14.(Eis.)/KG 55	Poltava	Oblt. Mathias Bermadinger	He 111	n.d.
I./KG 100	Poltava	Hptm. Hans-Georg Batcher	He 111	38
Stab/St.G 2	Krestovoy	Major Ernst Kupfer	Ju 87	3
I./St.G 2	Kharkov–East	Hptm. Bruno Dilley	Ju 87	37
II./St.G 2	Kharkov–North	Hptm. Hans-Karl Stepp	Ju 87	36
III./St.G 2	Kharkov–North	Hptm. Walter Krauss	Ju 87	35
Panzerjägerstaffel/St.G 2	Kharkov–East	Oblt. Helmut Schube	Ju 87	11
Stab/St.G 77	Tolokonnoye	Major Helmut Bruck	Ju 87	3
I./St.G 77	Tolokonnoye	Hptm. Karl Henze	Ju 87	40
II./St.G 77	Tolokonnoye	Hptm. Helmut Leicht	Ju 87	41
III./St.G 77	Kharkov–West	Major Georg Jakob	Ju 87	36
Stab/Sch.G 1	Bessonovka	Major Alfred Druschel	Fw 190	4
I./Sch.G 1	Bessonovka	Hptm. Georg Dorffel	Fw 190	52
II./Sch.G 1 (5. - 7. Staffeln)	Bessonovka	Hptm. Frank Neubert	Fw 190	49
			(7. Staffel = Hs 123)	(inc. 16 Hs 123)
Pz.Jagd-Kdo/Sch.G 1 (Führer der Panzerjäger)	Zaporozhye (to Varvarovka on 8 July 1943)	Hptm. Bruno Meyer	Hs 129	4
4.(Pz)Sch.G 1	Zaporozhye (to Varvarovka on 8 July 1943)	Oblt. Georg Dornemann	Hs 129	17
8.(Pz)Sch.G 1	Zaporozhye (to Varvarovka on 8 July 1943)	Hptm. Rudolf-Heinz Ruffer	Hs 129	16
4.(Pz)Sch.G 2	Zaporozhye (to Varvarovka on 8 July 1943)	Major Matuschek	Hs 129	17
8.(Pz)Sch.G 2	Zaporozhye (to Varvarovka on 8 July 1943)	Obit. Franz Oswald	Hs 129	10
Pänzerjagerstaffel/JG 51	Zaporozhye (to Varvarovka on 8 July 1943)	Obit. Hans Jentsch	Hs 129	15
Stab/JG 52	Bessonovka	OTL Dietrich Hrabak	Bf 109	4
I./JG 52	Bessonovka	Hptm. Johannes Wiese	Bf 109	34
II./JG 52	Kharkov–Rogan	Hptm. Gerhard Barkhorn	Bf 109	31
III./JG 52	Ugrim	Hptm. Gunther Rall	Bf 109	42
II./JG 3	Kharkov–Rogan and Ugrim	Major Wolfgang Ewald	Bf 109	33
III./JG 3	Bessonovka	Hptm. Heribert Rinke	Bf 109	40
NAGr 6	Veterinar			
1./NAGr 6 - 1.(H)/21			Fw 189	10
2./NAGr 16			Fw 189	10
1./NAGr 2-4.(H)/10	Kharkov		Bf 109	13
5.(H)/32			Hs 126	9
2.(H)/33			Bf 110	8
Transportstaffel Fl.K. VIII			Ju 52	13
Flugbereitschaft Fl.K. VIII			Fi 156	n.d.
San-Flugbereitschaft 3			Fi 156 and Ju 52	n.d.
Total number of aircraft				986

Units directly subordinated to *Luftflotte* 4 (General der Flieger Otto Dessloch)				
UNIT	AIRFIELD	COMMANDER	EQUIPMENT	STRENGTH
FAGr 4		Major Hans-Dietrich Klette		
1. Nachtaufklarungsstaffel (F)	Zaporozhye–South		Do 217 and He 111	12
2.(F)/22	Stalino		Ju 88	12
2.(F)/100			Ju 88	13
2.(F)/11	Kharkov		Ju 88	10
3.(F)/121			Ju 88	2
Wekusta 76	Zaporozhye		Ju 88	7
Storkampfgruppe Luftflotte 4				60 (total below)
1./ Storkampfgruppe Luftflotte 4	Kharkov–North		He 46, Do 17, Hs 126	
2./ Storkampfgruppe Luftflotte 4	Kharkov–North		Ar 66, Fw 189, Ju W 34	
3./ Storkampfgruppe Luftflotte 4	Kharkov–North		Go 145	
4./ Storkampfgruppe Luftflotte 4	Kharkov–North		Ar 66, Fw 58, Ju 87	
5./ Storkampfgruppe Luftflotte 4	Kharkov–North		He 46, Hs 126, Ar 66	
6./ Storkampfgruppe Luftflotte 4	Kharkov–North		Go 145	
Total number of aircraft				116

Soviet raids on German communications, supply lines and supply points around the salient were more effective – the Second and Sixteenth Air Armies flew 9896 sorties against these targets with some spectacular successes, such as the detonation of 1200 tons (1181 tonnes) of ammunition in the marshalling yards at Bryansk.

The repeated delays to the start date of Operation 'Citadel' allowed the *Luftwaffe* to undertake a limited strategic bombing campaign with a series of night attacks against Soviet war industries throughout much of June 1943. These raids were primarily directed against the GAZ tank factories in Gorkiy, the rubber industry at Yaroslavl and the oil refineries at Saratov. These provoked a series of Soviet retaliatory raids on *Luftwaffe* bomber airfields, but these inflicted only limited damage.

Soviet intelligence sources provided so much detail about German preparations for their offensive that the air armies had planned a series of pre-emptive strikes against *Luftwaffe* airfields as early as May. These plans had been regularly updated and it was confidently expected that the attacks would cripple German air power in the sector during the first crucial days of Operation 'Citadel'. Unfortunately for the VVS units involved, by the time that they were on their way to their targets, most of the *Luftwaffe* aircraft based at those airfields were already airborne in readiness to make their own attacks in support of the German offensive. A

further factor was the effective *Luftwaffe* radar coverage of this sector of the front – Freya radars at Belgorod and Kharkov detected the Soviet aircraft and directed fighters against them, which were able to break up many of the formations before they reached their targets.

The *Luftwaffe* was able to make devastating attacks against key sectors of the front. One Soviet account referred to the sector attacked by II SS Panzer Corps: '... an area measuring 6 x 4 kilometres in the sector of 52nd Guards Rifle Division ... was subjected to 15 hours of uninterrupted air attacks ... during this time as many as 20 bombers were continuously in the air. Such intense enemy air operations ... made it considerably easier for the Germans to penetrate our defence and advance 6–8 kilometres into the depth of our positions.' (The division sustained 8500 casualties on 5 July, the majority of which were due to these near-continuous air attacks.)

As the offensive went on, demands on the *Luftwaffe* mounted – in a typical action, Stukas were called in to silence a Soviet 120mm (4.77in) mortar battery, which was deployed in deep emplacements amongst a stand of poplar trees. The battery was virtually immune to German artillery fire as the shells detonated high up in the trees, but the dive-bombing did the job in a few minutes. II SS Panzer Corps was given a high priority for the available close air support and quickly developed 'standing operating procedures',

Hungarian units subordinated to *Fliegerkorps* VIII

UNIT	AIRFIELD	COMMANDER	EQUIPMENT	STRENGTH
1. Tavolfelderito osztaly		Ornagy Gyula Timar		
1/1.Tavolfelderito szazad	Kharkov–Osnova	Szazados Adorjan Mersich	Ju 88	
3/1. Kozelfelderito szazad	Kharkov–Grobly	Szazados Imre Telbisz	Fw 189	12
4/1. Bombazo szazad	Kharkov–Southeast	Szazados Tihamer Ghyczy	Ju 88	
5/1. Vadasz osztaly	Kharkov–South	Ornagy Aladar Heppes	Bf 109	
5/1. Vadasz szazad	Kharkov–Voychenko	Szazados Gyorgy Ujszaszy	Bf 109	
5/2. Vadasz szazad	Kharkov–Voychenko	Szazados Gyula Horvath	Bf 109	

which included directing Stuka formations against every copse or stand of trees along its line of advance to eliminate the anti-tank guns that they almost invariably contained.

Such intensive operations (six or more sorties per day were commonplace) took an increasing toll of Stuka units – not only was the Red Air Force fighter arm beginning to make its presence felt, but Soviet AA batteries were now deployed as part of the forward defences close to the anti-tank positions. This allowed the AA guns to protect their sector of the front against the *Luftwaffe* and to operate in the anti-tank role when necessary.

Besides such 'conventional' operations, the *Luftwaffe* was heavily committed to night harassment operations. These were inspired by the Red Air Force, which during 1942 had pioneered the use of obsolete aircraft such as the Polikarpov Po-2 in night harassment raids against the German rear areas. Noting the success of these raids, the *Luftwaffe* quickly copied the technique. In December 1942, the first *Störkampfstaffel* (harassment squadron) was formed and equipped with a mixture of Gotha Go 145 and Arado Ar 66 aircraft drawn from flying training schools. This unit was so effective that similar squadrons were formed across the Eastern Front, equipped with an amazing variety of obsolete aircraft. By mid-1943, *Luftflotte* 4 fielded six such squadrons with a total of 60 assorted aircraft, all of which were based at Kharkov-North:

1. *Störkampfstaffel* He 46/Do 17/Hs 126
2. *Störkampfstaffel* Ar 66/Fw 189/Junkers W 34
3. *Störkampfstaffel* Go 145
4. *Störkampfstaffel* Ar 66/Fw 58/Ju 87B/R/D
5. *Störkampfstaffel* He 46/Hs 126/Ar 66
6. *Störkampfstaffel* Go 145

The biplane types that equipped many of these units copied the tactics of their Soviet counterparts: flying only a few metres above the ground, climbing for the final approach, then cutting the engine and making a gliding bombing run. This had the effect that the targeted troops were left with only the eerie whistling of the wind in the wings' bracing-wires as an indication of the impending attack. Their average maximum load of 6 x 50kg (110lb) bombs was unlikely to inflict major damage on any given target, but the disruption caused by constant harassment raids night after night steadily sapped the combat capability of even the best units.

LUFTWAFFE SUCCESSES

One of the most spectacular *Luftwaffe* attacks was launched on 8 July against II Guards Tank Corps, which had been ordered to attack the right flank of II SS Panzer Corps in an attempt to relieve the pressure on Sixth Guards and First Tank Army. The primary objective was to cut the Belgorod–Orel road, which was the main German supply route for this sector of the front. This seemed to be a promising operation because *Totenkopf* was in the process of handing over the defence of the area to the newly arrived 167th Infantry Division. Moreover, Soviet operational security had been very good and the Germans did not suspect that II Guards Tank Corps was in the area. All this careful preparation was, however, ruined by pure chance – a flight of Hs 129 'tank busters' led by *Hauptmann* Bruno Meyer, the commander of *Panzerjagdkommando Weiss,* sighted the corps' lead elements. Meyer radioed a sighting report to the Kommando's base at Mikoyanovka and the unit's four *Staffeln,* each with 16 Henschels, took off in relays to keep II Guards Tank Corps under constant attack. They were joined by the Fw 190s of Major Druschel's *Schlachtgeschwader* 1 in a textbook display of air power. The Fw 190s targeted the AA guns and infantry with SD-2 cluster bombs and strafing runs, allowing the Hs 129s to attack the Soviet tanks with their 30mm MK 103 cannon. The tungsten-cored 30mm (1.18in) APCR rounds easily penetrated the engine decks and rear armour of the T-34s and T-70s, and the *Luftwaffe* claimed a total of at least 50 tanks destroyed in three hours of attacks, which managed to halt II Guards Tank Corps' without the involvement of any German ground forces.

Soviet AA guns were generally regarded as the *Luftwaffe's* most formidable adversary, both with respect to numbers and quality. Every German formation came under heavy fire as soon as it flew over the Red Army's front line – Soviet 25mm (0.98in) and 37mm (1.45in) light AA guns had gained a reputation for accuracy, although the heavy 76.2mm (3in) and 85mm (3.35in) weapons were handicapped by primitive fire control equipment. Even if they avoided light AA batteries, low-flying aircraft were constantly at risk from rifle and machine-gun fire from all Red Army units within range. Although this was generally inaccurate, the

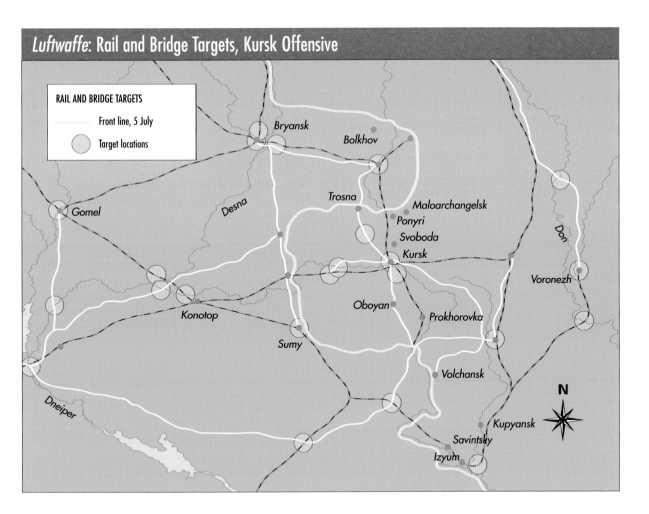

Luftwaffe: Rail and Bridge Targets, Kursk Offensive

RAIL AND BRIDGE TARGETS

.... Front line, 5 July

◯ Target locations

Bryansk
Bolkhov
Gomel
Desna
Trosna
Maloarchangelsk
Ponyri
Svoboda
Kursk
Voronezh
Don
Konotop
Oboyan
Prokhorovka
Sumy
Volchansk
N
Dnieper
Kupyansk
Savintsky
Izyum

RAIL AND BRIDGE TARGETS
The appalling state of Russian roads meant that railway lines were vitally important in ensuring the rapid reinforcement and resupply of the forces defending the Kursk salient. The tracks, marshalling yards and railway bridges were high-value targets for the *Luftwaffe*, but were heavily defended and difficult to hit. Even when bombing succeeded in causing significant damage, huge numbers of civilians were conscripted by the NKVD and forced to work around the clock to carry out repairs.

sheer volume of fire posed a significant threat to aircraft without good armour protection. On the whole, German pilots considered Soviet AA units to be far more dangerous than their fighters.

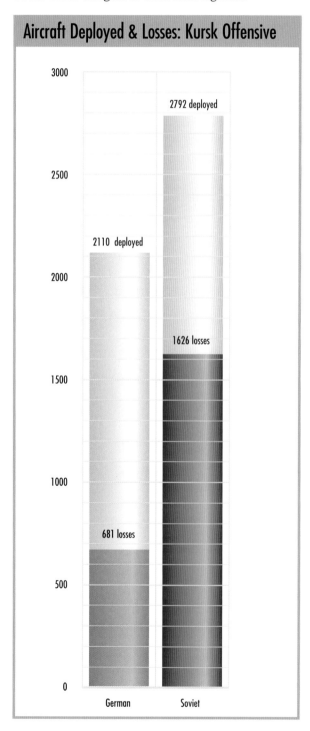

Aircraft Deployed & Losses: Kursk Offensive

3000

2792 deployed

2500

2110 deployed

2000

1626 losses

1500

1000

681 losses

500

0

German   Soviet

The *Luftwaffe's* quick and effective close air support in the southern sector of the Kursk salient depended on smoothly functioning communications between the attacking ground formations and HQ *Fliegerkorps* VIII, which had operational control of the bulk of *Luftflotte* 4's combat aircraft operating over this sector of the front. The *Luftwaffe* had maintained a corps of liaison officers since the beginning of the war, drawn from aircrew experienced in close air support operations. As usual during an offensive, these liaison teams were attached to Army Group South's corps and divisional HQs. They acted as dive-bomber and fighter controllers, using their radios to direct approaching formations to the targets indicated by the ground commander, to correct their fire, and to provide updates on the current tactical air situation in the area. These liaison teams also reported directly to *Fliegerkorps* VIII every two hours with details of the ground situation, air situation and any other relevant information. This procedure kept both the controlling HQ and the air formations constantly informed regarding the progress of the attack.

Although most accounts of air warfare over the Kursk salient understandably concentrate on the relatively glamorous aspects of fighter combat and ground attack, the *Luftwaffe* could not have fought the campaign without its over-worked and increasingly vulnerable air transport squadrons. Their Ju 52s ferried equipment and critical supplies to the operational airfields and evacuated the wounded on their return flights, supplementing the efforts of the dedicated medical evacuation aircraft. The appalling road conditions slowed the delivery of all types of supplies and meant that urgently needed spare parts frequently had to be airlifted so that the formations spearheading the advance could keep moving. Liaison squadrons were also indispensable, transmitting orders and delivering maps, messages and aerial photographs. On many occasions these liaison aircraft also evacuated seriously wounded personnel on their return trips.

LOSSES COMPARED: Although the Red Air Force was rapidly gaining strength by mid 1943, it was still grossly inferior to the *Luftwaffe* in terms of aircrew training standards. This discrepancy is reflected in the losses shown. (A further factor was the poor quality control in Soviet aircraft factories, which contributed to both combat losses and an appalling accident rate.)

Ground Attack Aircraft

The *Luftwaffe* deployed a remarkable variety of aircraft in the ground attack role over the Kursk salient, ranging from the most modern Fw 190s to the ancient biplanes of the *Störkampfstaffeln.*

A full study of all the *Luftwaffe's* ground attack aircraft would require a book to itself, but this section can at least provide a summary of the principal types and their main weapons.

FOCKE-WULF FW 190F

The Fw 190 entered service with the *Luftwaffe* as a fighter in September 1941 and soon proved to be superior to most of its Allied opponents. Unsurprisingly, fighter-bomber variants were quickly developed which were as formidable as their interceptor counterparts. In many respects, the Fw 190 was an ideal ground attack aircraft – its radial engine was far less vulnerable to ground fire than the in-line liquid-cooled engines of many other contemporary fighters and it had outstanding low-altitude performance. Most fighter-bomber variants had substantial armour on their undersides, plus a 50mm (1.97in) armoured glass windscreen and an 8mm (0.32in) armoured seat incorporating 13mm (0.51in) head and shoulder armour, which afforded

During Operation 'Citadel', the Luftwaffe attempted to recreate the successes of earlier years by using Stukas as flying artillery.

their pilots a considerable degree of protection from enemy ground fire.

By mid-1943, the Fw 190F was fast becoming the standard fighter-bomber variant on the Eastern Front – this was essentially an adapted Fw 190A fighter with substantial armour plate under the engine and cockpit for protection against ground fire, stronger landing gear to support greater take-off loads, and other modifications. Although performance was reduced by these changes, all the fighter-bomber variants were still dangerous adversaries in air combat and scored a large number of kills. The Fw 190F-1 sub-variant was based on the FW-190A-4. Only a small number were built, for evaluation purposes. The F-1 had reduced gun armament, with two MG17 7.92mm (0.31in) machine-guns in the cowling and an MG151 20mm (0.79in) cannon in each wing. This allowed it to be fitted with racks under the fuselage for the carriage of one 500kg (1100lb) or four 50kg (110lb) bombs, plus an optional rack under each wing for a single 250kg (550lb) bomb or two 50kg (110lb) bombs.

The Fw 190F-2 was a derivative of the A-5 and featured a new bulged canopy to improve pilot

vision, with the first of this series coming off the production line in early 1943. The Fw 190F-3 was similarly derived from the A-6, with the first deliveries being made in the summer of 1943.

HENSCHEL HS 129

The Hs 129 originated from an RLM requirement for a twin-engine ground attack aircraft issued in April 1937. This specified a well-armoured design with an armament of at least two 20mm (0.79in) cannon, coupled with a proviso that the type should be powered by low-powered engines that were not required for other designs. The intention was to make full use of all available engines, but the effect was to ensure that the design was badly under-powered and to cripple its development potential.

The Hs 129 was designed around a single large 'bathtub' of armour plate that made up the entire nose section of the aircraft, completely enclosing the pilot up to head level. Even the canopy was largely armour plate, with an appalling field of vision for the pilot through a small windscreen and

tiny side windows. The near-triangular cross-section of the fuselage was extremely strong, but resulted in a very cramped cockpit, which was so small that some of the engine instruments were fitted on the engine nacelles and the gun sight was mounted externally on the nose.

The prototype Hs 129 first flew in May 1939 and production of the A-series began just over a year later. These were fitted with 347kW (465hp) Argus As 410 engines and it rapidly became apparent that they were so grossly under-powered that they were unfit for combat. Urgent design studies were made to assess the feasibility of fitting alternative engines, and it was decided to use 522kW (700hp) Gnome-Rhône 14M radial engines, which made the type a viable combat aircraft, although it remained markedly under-powered.

Only a few A-series aircraft were completed before production switched to the Gnome-Rhône powered Hs 129B-1 in December 1941. These were fitted with a new windscreen and cockpit canopy with far more extensive glazing, which dramatically improved the pilot's field of vision.

Focke-Wulf Fw 190F-2

The Fw 190F series were dedicated ground attack variants of the basic Fw 190A-series fighters fitted with additional armour, which entered service in early 1943. Despite the extra weight of this armour, they were still capable of effectively defending themselves against Soviet fighters and were far less vulnerable than earlier *Luftwaffe* ground attack aircraft.

■ CREW

■ SPECIFICATIONS

Type: Single-seat ground attack aircraft
Length: 8.95m (29ft 7in)
Wingspan: 10.51m (34ft 5in)
Height: 3.95m (12ft 11in)
Wing area: 18.3m² (196.99ft²)
Empty weight: 3325kg (7330lb)
Loaded weight: 4417kg (9735lb)

Max takeoff weight: 4900kg (10,800lb)
Powerplant: 1 x 1250kW (1700hp) BMW 801 D-2 radial engine
Maximum speed: 635km/h (395mph)
Range: 750km (466 miles)
Service ceiling: 10,600m (34,777ft)
Rate of climb: 10.7m/s (2106ft/min)

Wing loading: 241kg/m² (49.4lb/ft²)
Armament: 2 x 13mm (0.51in) MG131 machine-guns (475rpg) plus 2 x 20mm (0.79in) MG151/20 cannon (250rpg); bomb load of up to 1800kg (3968lb)

The type became operational on the Eastern Front in May 1942, shortly after which units began to receive modification kits for fitting the 30mm (1.18in) MK101 cannon. Each aircraft carried a single MK 101 fed by a 30-round drum magazine in an under-fuselage gun pack and the weapon's tungsten-cored APCR rounds proved to be capable of penetrating the side and rear armour of T-34s and all Soviet light tanks.

By the time of Operation 'Citadel', most frontline units had re-equipped with the Hs 129B-2. This variant incorporated numerous minor improvements besides replacing the MK 101 with the lighter, faster-firing and more compact belt-fed 30mm MK103 cannon, which had a similar armour-piercing capability. These aircraft proved to be highly effective in their intended role when operating over the Kursk salient, but their sluggish performance made later operations extremely hazardous as the Red Air Force gained an increasing degree of air superiority.

STUKA! JUNKERS JU 87

The *Luftwaffe* had relied heavily on the precision-bombing capability of the Stuka since the beginning of the war. The type had played a vital role in Russia from the very first days of Operation *Barbarossa*, and even in mid-1943 it was still a potent weapon when operating in conditions of *Luftwaffe* air supremacy.

By this time, the vast majority of front line units were equipped with variants of the Ju 87D. The D-series originated as a desperate move to update the design pending the development of an entirely new dive-bomber, a move made necessary because the Battle of Britain had shown that earlier variants could not operate successfully when opposed by modern fighters.

The initial D-1 model was powered by a 1045kW (1400hp) Jumo 211J-1 engine, driving a new VS-11 propeller. The new engine permitted a much cleaner installation than its predecessors, and the airframe was redesigned accordingly with a new engine cooling system, eliminating the older 'broken nose' appearance of the Stuka. An entirely new streamlined canopy was fitted, whilst the main landing gear spats were reduced in size and tidied up. These spats were often removed by frontline units, as they didn't offer much improvement in

speed and were a nuisance in muddy field operations. Greater engine power also permitted more armour and fuel capacity, with the Ju 87D-1 featuring the outer wing tanks pioneered by the Ju 87R series.

The D-1 retained the twin fixed forward-firing 7.92mm MG 17 machine guns, but replaced the rear gunner's single 7.92mm (0.31in) MG15 gun with a twin-barrelled 7.92mm (0.31in) MG81Z, which had a better rate of fire.

The Ju 87D-1 had a strengthened airframe and fuselage bomb crutch, which allowed it to carry a 1800kg (3970lb) bomb over short ranges. A more typical bombload was a 1000kg (2200lb) bomb on the crutch and two 50kg (110lb) bombs under each wing, though when used in the close support role the wing racks were usually fitted with various *Waffenbehaelter* (weapons containers). The most common versions of these were fitted with either six MG81 machine guns, or twin 20mm (0.79in) MG-FF cannon.

The Ju 87D-1 began to replace the Ju 87B-2 in production in mid-1942. A similar D-2 variant was built in parallel, differing only in having a strengthened rear fuselage and a stronger tail wheel leg with a glider towing attachment. Towards the end of 1942, the D-3 was introduced – this was fitted with extra armour for the close support role. The steadily increasing load of armour and weapons was by now seriously affecting the Stuka's handling characteristics and the next sub-type, the D-5, was fitted with extended wingtips to cope with the additional weight.

The final major variant to see action over the Kursk salient was the Ju 87G anti-tank aircraft armed with two of the new 37mm (1.45in) Bordkanone BK 3.7 in large under-wing gun pods. Although these severely restricted the aircraft's manoeuvrability, there was no doubt about the effectiveness of the 37mm (1.45in) – using tungsten-cored *Hartkernmunition* APCR rounds, it could penetrate at least 70mm (2.75in) of armour at close range, sufficient to deal with almost all Soviet AFVs.

Hans-Ulrich Rudel flew the only production Ju 87G thought to have seen action during Operation 'Citadel', although a significant number of Ju 87Ds were fitted with the 37mm (1.45in) cannon, and operated as 'interim' Ju 87Gs during this period.

LUFTWAFFE BOMBS

The *Luftwaffe* made extensive use of fragmentation bombs throughout the war. Initially, the most common bombs in this category were the 10kg (22lb) SD-10 and 50kg (110lb) SD-50. Whilst these were satisfactory weapons, they required accurate delivery and it was rapidly appreciated that there was a need for an area-saturation fragmentation weapon.

The *Sprengbombe Dickwandig* 2kg or SD-2 (butterfly bomb) was quickly developed and entered service in 1940. They were dropped in dispensers holding up to 108 bombs, which opened just after release, scattering the bombs over a very wide area.

The SD-2 was an 8cm (3.2in) cylinder of cast iron, fitted with a 15cm (6in) steel cable that was attached via a spindle to the fuse screwed into the side of the bomb. The outer shell would hinge open as two half-cylinders when it was dropped, and spring-loaded vanes at the ends would flip out. These rotated in the airflow, turning the spindle as the bomb fell to arm the fuse. The SD-2 contained 225g (8oz) of TNT. They were generally lethal within a radius of 25m (82ft) and could inflict serious splinter injuries (such as penetrating eye wounds) as far away as 100m (330ft).

The SD-2 could be fitted with a variety of fuses, the principal types being:

- 41 fuse – impact fuse
- 67 fuse – clockwork time delay, adjustable from between 5 and 30 minutes after impact
- 70 fuse – anti-handling device, detonating if the bomb was moved after impact

Junkers Ju 87G 'Stuka'

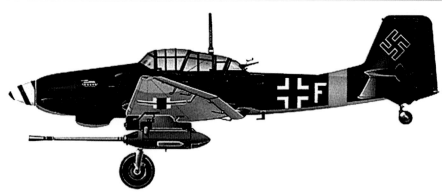

Problems with the early Ju 88P and Hs 129B anti-tank aircraft led to the decision to develop a 'tank-buster' variant of the Stuka in November 1942. Initial studies envisaged an armament of two 30mm (1.18in) cannon plus a bomb load of up to 1000kg (2200lb). However, at the suggestion of the Stuka ace Hans-Ulrich Rudel, the prototype was fitted with two adapted 37mm (1.45in) Flak 18 guns, designated Bordkanone BK 3.7, in under-wing gun pods. This machine first flew on 31 January 1943 and the type was adopted as the Ju 87G ('Gustav the tank killer'). The first production Ju 87G-1s were delivered to frontline units in April 1943 and proved to be devastatingly effective tank killers during the Kursk offensive, although they were extremely vulnerable to Soviet fighters.

CREW

SPECIFICATIONS

Type: Two-seat anti-tank aircraft
Length: 11.5m (37ft 8.75in)
Wingspan: 13.8m (45ft 3.5in)
Height: 3.9m (11ft 9.5in)
Wing Area: 31.9m² (343.38ft²)
Empty weight: 3900kg (8598lb)
Loaded weight: 5720kg (12,600lb)

Max takeoff weight: 6600kg (14,550lb)
Powerplant: 1 x 1037kW (1400hp) Junkers Jumo 211 J-1 12-cylinder inverted-vee engine
Maximum speed: 344km/h (214mph)
Range: 790km (491 miles)
Service ceiling: 7290m (23,915ft)

Rate of climb: n/k
Wing loading: 179.31kg/m² (36.69lb/ft²)
Armament: 2 x 37mm (1.45in) fixed forward-firing BK 37 cannon plus twin MG81Z 7.92mm (0.31in) machine guns in dorsal position.

Henschel Hs 129B-2

The requirement for a dedicated ground attack aircraft originated with the Condor Legion's reports on the Spanish Civil War. The prototype Hs 129 first flew in May 1939 and was accepted for service despite very poor performance due to its heavy armour and under-powered Argus engines. The adoption of more powerful Gnome-Rhone radial engines after the conquest of France, combined with much detailed redesign, made the type a viable combat aircraft and the first Hs129B-1s entered service in April 1942. A further range of improvements were incorporated in the Hs 129 B-2, which formed the great majority of the 865 aircraft completed by the time that production ended in September 1944.

■ **CREW**

■ **SPECIFICATIONS**

Type: Single-seat anti-tank and close-support aircraft
Length: 9.75m (31ft 11.75in)
Wingspan: 14.2m (46ft 7in)
Height: 3.25m (10ft 8in)
Wing area: 29m² (312.16ft²)
Empty weight: 3810kg (8,400lb)
Loaded weight: 5110kg (11,265lb)

Max takeoff weight: 5250kg (11,574lb)
Powerplant: 2 x 522kW (700hp)
  Gnome-Rhone 14-cylinder radial engines
Maximum speed: 407km/h (253mph)
Range: 688km (427 miles)
Service ceiling: 9000m (29,525ft)
Rate of climb: 7.083 m/s (1394 ft/min)
Wing loading: 176.21kg/m² (36.09lb/ft²)

Armament (All fixed forward-firing):
  2 x 20mm (0.79in) MG151/20 cannon
  (125rpg), 2 x 13mm (0.51in) MG131
  machine-guns (250rpg); in addition, a
  single 30mm (1.18in) MK101 cannon
  could be carried in a pod under the
  fuselage. Later sub-types carried a single
  37mm (1.45in) BK37 cannon.

The Germans forbade the use of SD-2s with anti-disturbance fuses against retreating opponents due to the risks to friendly forces. They were intended for use against targets behind enemy lines, for harassing effect only. The Germans fully appreciated the value of these little bombs against troop movements. Colonel Lovell, a member of a British military mission to the Soviet Union, who advised the Red Army on bomb disposal matters, found that the Russians attached the greatest importance to the type:

'Used in high concentrations it had cost the Red Army great numbers of casualties and effectively held up the movement of formations. Russian soldiers had been reduced to detonating bombs by rifle fire, a method certain to cause casualties since the butterfly's fragmentation range was a hundred yards, at which distance it presented, at best, a poor

target – and the rifleman was bound to have his face toward the bomb.'

Although the SD-2 proved to be devastatingly effective against Soviet infantry, anti-tank and artillery positions during Operation 'Citadel', it was ineffective against AFVs. In the early stages of Operation *Barbarossa*, the *Luftwaffe* had used conventional bombs against armoured units, but nearly all these required direct hits to ensure a 'kill'. By mid-1943, the problem had been solved by the 4kg (8.8lb) SD-4HL bomb, which had a hollow charge warhead that could penetrate up to 125mm (5in) of armour, sufficient to destroy even the heaviest Soviet AFVs. These were also dropped from dispensers – 74 bombs were carried in the AB 500-1, whilst the AB 250 held 40 bombs. (Larger 550lb hollow charge bombs were used against bunkers, penetrating up to 3m/10ft of ferro-concrete.)

Chapter 4
The Test of Battle: Prokhorovka

In the spring of 1943, the fate of the Eastern Front hung in the balance – after achieving a major victory when the last remnants of Paulus' Sixth Army surrendered at Stalingrad on 2 February, the Red Army had overrun a great swathe of the eastern Ukraine and seemed poised to destroy Army Group South.

However, Stalin allowed himself to be carried away by success and pushed the offensive too far, leaving his forces grossly over-extended. *Generalfeldmarschall* von Manstein was not one to ignore such an opportunity and launched a devastating 'backhand offensive' spearheaded by the SS Panzer Corps, which badly mauled the Red Army – the Southwestern Front lost 23,000 men, 615 AFVs and 354 guns, whilst the Voronezh Front's casualties were even worse, totalling 40,000 men, 600 tanks and 500 guns. German forces once again held much of the territory lost during the winter except for a large salient centred on the small provincial city of Kursk.

OPPOSITE: Panzergrenadiers of the *Das Reich* division rest in a captured Soviet defensive position. All are wearing the summer pattern SS camouflage smock. The senior officer present is standing in the trench; he is an *Untersturmführer* (second lieutenant), with three pips on his tunic collar.

The Military Balance

As the spring thaw set in, halting all large-scale military operations, both sides were temporarily exhausted and grateful for the opportunity to build up their battered forces for the summer's campaign. Both Hitler and Stalin pored over the various options presented by their planning staffs, knowing that the next few months would have a decisive impact on the outcome of the war.

By the spring of 1943, there was no doubt about Soviet numerical superiority on the Eastern Front as a whole. Comparative strengths of the two sides were:

■ Axis – 2,700,000 men, 2209 AFVs and 6360 guns
■ Soviet – 6,000,000 men, 15,000 AFVs and 33,000 guns.

Despite the massive losses sustained by Axis forces since the opening of Operation *Barbarossa,* they were actually winning the battle of attrition. Although Soviet reserves of manpower were easily double those of Germany and her allies, Red Army casualties were several times greater than this in 1941/42 and even in 1943; the casualty ratio was 4:1 in favour of the Axis. It was a similar story as far as AFVs were concerned – Soviet tank production was more than twice that of German factories, but the loss ratio stood at more than 3:1 in the Germans' favour.

This was an indication of just what elite formations such as *Das Reich* could achieve – notably its Tiger company, which was establishing a kill ratio of better than 8:1. (By the end of the war, it would have destroyed a total of 250 AFVs for the loss of 31 Tigers.)

WHY KURSK?

The Kursk salient was very much unfinished business after von Manstein's counter-offensive. (He had fully intended to eliminate the salient in March, but Soviet reinforcements, the spring thaw and the exhaustion of his forces had compelled him to halt operations before this could be achieved.) To the Germans, the salient measuring 200km (120 miles) from north to south with a depth of up to 150km (90 miles) was a potential 'springboard for the Soviet reconquest of the Ukraine'. Something would clearly have to be done to eliminate the threat that it posed, but there was initial uncertainty over just what that something should be.

As early as February 1943, von Manstein had proposed a 'backhand' offensive for the coming summer – this was essentially a large-scale version of his successful winter counter-offensive. His plan envisaged concentrating a powerful armoured reserve around Kiev before luring the Red Army into a series of advances deep into the Ukraine as far west as the Dneiper. The panzers would then be unleashed in a massive attack with the aim of trapping and destroying the Soviet spearheads between the Sea of Azov and the Dnieper.

As von Manstein put it:

'… if the Russians did as we anticipated and launched a pincer attack on the Donets area from the north and south, an operation which would sooner or later be supplemented by an offensive around Kharkov, our arc of front along the Donets and Mius should be given up in accordance with an agreed timetable in order to draw the enemy westwards towards the Lower Dnieper. Simultaneously, all the reserves that could possibly be released, including the bulk of the armour, were to assemble [in the vicinity of Kiev], first to smash the enemy assault forces which we expected to find there and then to drive into the flank of those advancing in the direction of the Lower Dnieper. In this way the enemy would be doomed to suffer the same fate on the Sea of Azov as he had in store for us on the Black Sea.'

OPPOSITE: The elimination of the Kursk salient was the obvious objective for the main German summer offensive of 1943. With his usual flair for achieving strategic surprise, von Manstein proposed a gigantic 'backhand offensive' (A). As a preliminary, the Red Army would be lured into a headlong advance westwards towards the Lower Dneiper by an exaggerated show of German weakness. As soon as Soviet forces were fully committed and overstretched, he intended to launch a massive counter-attack by a powerful armoured force secretly assembled near Kiev to cut off and destroy the Russian spearheads. It was a bold and imaginative plan – too bold and unorthodox for Hitler who insisted on the seemingly safer and simpler 'forehand offensive', Operation 'Citadel' (B).

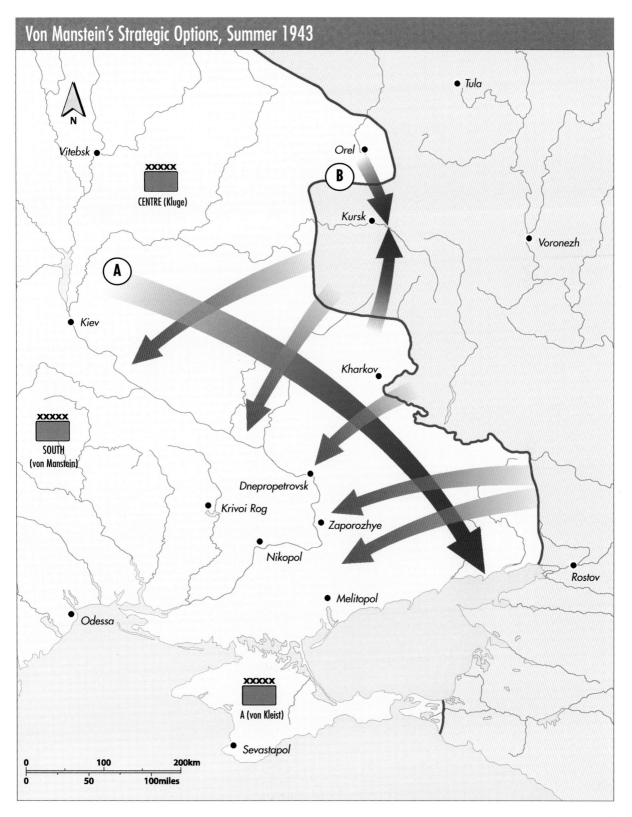

Von Manstein's Strategic Options, Summer 1943

This plan had the potential to inflict a stunning defeat on the Red Army, but it was too unorthodox for Hitler, who instinctively hated the thought of voluntarily surrendering any territory. He was also suspicious of any plan that would give such wide-ranging freedom of action to a single commander in contravention of his aim of exercising the greatest possible control of his generals' conduct of the war.

TRADITIONAL PINCER MOVE

Unsurprisingly, Hitler rejected the plan, opting for the much more conventional 'forehand' offensive – pincer attacks from north and south to surround and destroy the forces defending the salient which were

A Panzer Mk VI Ausf E Tiger tank from the *Das Reich* division deploys at the beginning of the Kursk offensive, identifiable from its two vertical upright strokes on the hull of the tank.

estimated at 60 divisions and five or six tank corps. It was confidently expected that their loss would cripple the Red Army and yield up to 700,000 prisoners – potentially invaluable slave labour for German war industries. (The elimination of the salient would also significantly shorten the German front line, releasing up to 20 divisions to reinforce other key sectors.)

The planning for the offensive received top priority following the issue of Hitler's Operational Order No. 6 on 15 April – the definitive directive for the Kursk offensive, designated *Zitadelle* ('Citadel'). This stressed the urgency of the operation, which was scheduled for 3 May, stating that it 'must succeed rapidly' – this urgency was partly due to the military risks of the offensive, but also recognized the political imperative to restore the confidence of Germany's increasingly reluctant Italian, Rumanian and Finnish allies.

As Operational Order No. 6 was being drafted, Stalin was reluctantly deciding that he had to ignore his instinctive preference for a pre-emptive offensive and follow his commanders' advice not to attack until the Germans had weakened themselves in assaults against properly prepared Soviet defences in the Kursk salient. The question was how much time the Red Army would be given to prepare those defences.

WHY PROKHOROVKA?

Although the German offensive was a two-pronged attack, the south was the *Schwerpunkt* – the focal point of the operation. The assault was to be spearheaded by Colonel General Hoth's Fourth Panzer Army of two panzer corps fielding a total of 1100 tanks and assault guns, the most powerful German armoured force to be assembled during the entire war. (These were XLVIII Panzer Corps and II SS Panzer Corps, the latter comprising *Leibstandarte*, *Das Reich* and *Totenkopf*.)

Operational Order No. 6 had stated that the attack in this sector would be made '... with strongly concentrated forces from the Belgorod–Tomarovka line, break through the Prilepy–Oboyan line and link up with the attacking armies of Army Group Centre east of Kursk'.

This was the direct route, but when von Manstein and Hoth began detailed operational planning, it became clear that it was unlikely to be the quickest or the surest way of securing its objectives. They identified the following drawbacks to this route:

■ The terrain south of Oboyan was poor tank country – much of the area was marshy and strewn with minor water obstacles in addition to the River Psel, which would have to be crossed in the teeth of strong Soviet defences.

RIGHT: Kursk was such an obvious objective that the Soviets began fortifying it almost as soon as the Germans decided to attack it. As early as March, Marshal Georgi Zhukov and his Front commanders were presenting Stalin with their expectations of likely German plans for the coming campaigning season. Their predictions proved to be remarkably accurate when the battle started in July.

In addition, the Red Army planned new offensives of its own, scheduled to open the moment the German attack stalled. Stalin and his most senior commanders gambled that they could hold Kursk against the elite Panzer divisions, absorb the full strength of the German blow, then unleash a multi-Front offensive that would liberate the Ukraine.

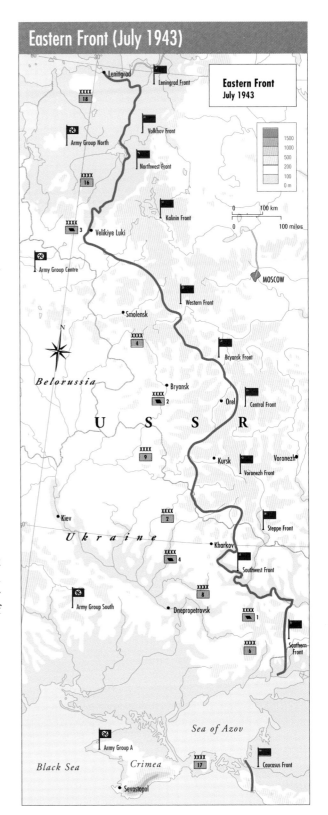

Eastern Front (July 1943)

Eastern Front
July 1943

147

■ The importance of Oboyan was so obvious that it was almost certain to be one of the main concentration points for Red Army armoured formations being brought in to reinforce the salient.
■ It was believed that the main Soviet strategic armoured reserve (including Fifth Guards Tank Army) was assembling around Korocha and that it would intervene as soon as Fourth Panzer Army was committed to fighting its way across the Psel. The obvious Russian approach route was across the land bridge between the rivers Psel and Donets, through the town of Prokhorovka. There was real risk that following Operational Order No. 6 would allow a mass of Soviet armour to smash Hoth's right flank as he was struggling to force a crossing of the River Psel, creating a situation which he felt could 'quickly turn into a disaster'.

Hoth therefore proposed that on approaching the Psel, Fourth Panzer Army should change its line of advance and swing northeast to intercept the Soviet armoured reserves around Prokhorovka. (Another important factor in the equation was III Panzer Corps, which was to spearhead the advance of Army Detachment Kempf covering Hoth's right flank. Its line of advance should bring it into action just south of Prokhorovka, with a good chance of trapping the Russian armour between the two forces.) After eliminating this threat, the German forces would again swing north, cross the Psel east of Obayan and resume the drive on Kursk. The proposal was readily accepted by von Manstein and formed the basis of all future operational plans for this sector.

Das Reich Prepares for the Offensive

The division had lost 102 officers and 4396 other ranks killed and wounded in action during the period from 30 January to 20 March – almost 25 per cent of its total strength.

The stabilization of the front at the end of March allowed *Das Reich* to spend the next two months reconstituting its battered units and training the influx of reinforcements.

Some of these were veterans who had been sent back to Germany after being wounded in the battles of February and March. Many of them cut short their convalescent leave and made their own way back to the division to avoid being posted to training units. Despite the fury of the *Waffen-SS Personalamt* (Personnel Department), which bombarded *Das Reich's* staff with enquiries about the missing convalescents, the division looked after its own and refused point-blank to return the 'fugitives'.

Whilst the returning veterans brought welcome expertise, many of other replacements were new recruits who were inadequately prepared for the realities of combat on the Eastern Front; the intensive pre-war *Waffen-SS* training was by now a rapidly fading memory. This had included several months of intensive basic training concentrating on physical fitness, small-arms proficiency and political

indoctrination. The one in three trainees who passed the course were then sent to specialist schools, where they received further training in their chosen combat arm.

In the spring of 1943, training within the division had to be far more basic, primarily covering the essentials of combat and survival in the uniquely harsh conditions of Russia. Live firing training figured largely in the programme – the inevitable injuries and fatalities being accepted because it was believed that this saved far heavier casualties resulting from raw troops 'freezing' when coming under fire for the first time.

Whilst units were training their reinforcements, the division's engineers were busy near the front line using air reconnaissance photographs to construct a full-sized replica of the Soviet defences which it would have to attack. Elements of the division were able to use this replica to hone their assault tactics in a series of exercises. (The most impressive of these was held on 5 June, when the entire division carried out a live-firing exercise supported by heavy artillery and the Ju 87s of Stuka *Geschwader* 77.)

BELOW: Each Soviet defence line was a deep complex of trenches, minefields, barbed wire entanglements and anti-tank positions designed to slow the progress of German armoured attacks. The expectation was that such defences would severely weaken the German assault forces, which could then be destroyed in a series of counter-attacks by the armoured reserves of the Central and Voronezh Fronts. These counter-attacks were intended to pave the way for the major Soviet offensives to take Orel, Belgorod and Kharkov. The defence of the northern sector of the salient was successful enough to allow the offensive against Orel to be launched as planned. In contrast, the attacks on Belgorod and Kharkov were delayed by the alarming progress of Fourth Panzer Army in the south, which was contained only by drawing on Fifth Guards Tank Army and other reserves intended for these offensives.

THE SOVIET 'CITADEL'

The Red Army began to build field defences in the Kursk salient as soon as the front lines stabilized in late March 1943, but these were insignificant in comparison with the massive construction programme that began in April as soon as the decision had been taken to await the German offensive rather than attempt a pre-emptive strike. Eight defensive zones were completed with a total depth of roughly 160km (100 miles). Their positioning was carefully selected to cover all eventualities – only five were built in the salient

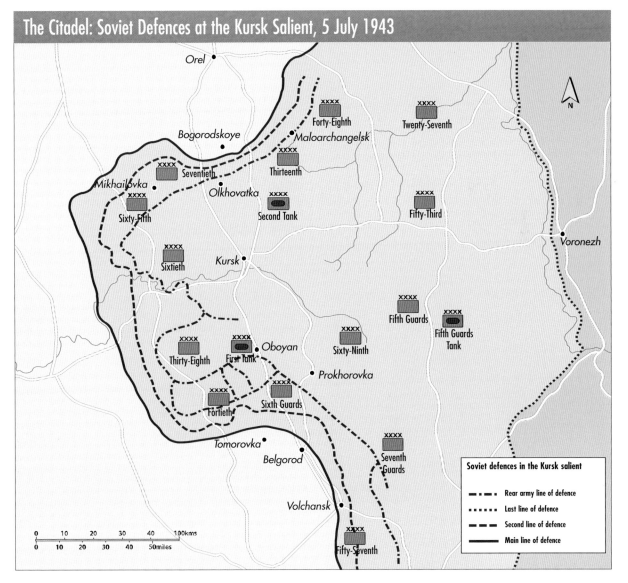

The Citadel: Soviet Defences at the Kursk Salient, 5 July 1943

Soviet defences in the Kursk salient

- ▬ ·▬ ·▬ · Rear army line of defence
- · · · · · · Last line of defence
- ▬ ▬ ▬ Second line of defence
- ▬▬▬ Main line of defence

itself; the remainder were constructed across its 'neck' so that the area could be sealed off if the Germans succeeded in breaking through to Kursk.

The scale of the construction work was staggering – whilst the first two defence lines were built by military units, the remainder were largely completed by a workforce of impressed local civilians, which may well have totalled 300,000. (A high proportion of these were women because most remaining men of military age were hastily conscripted to bring the units defending the salient up to strength.)

The defences were optimized for the anti-tank role because it was recognized that the German offensive would almost certainly rely on armoured strength to compensate for the heavy losses that had seriously weakened their infantry formations. The forward edge of each defence line was a maze of barbed wire entanglements, anti-tank obstacles and minefields, all of which were covered by artillery fire.

MINES

The Red Army laid a total of 640,000 mines of all types in defence of the salient. These were positioned in front of the main defence lines and in depth to channel attacking armour into 'killing zones'. The commander of the Sixty-Fifth Army recalled that they posed a real risk to the defenders because they 'stretched from the front line far into the rear. … They were so extensive that we had to post warning pickets and mark the mined areas with signs.' Throughout the offensive, the main minefields were supplemented by mobile mine-laying detachments, which could rapidly be deployed to any threatened sector.

The density of the minefields was well in excess of anything previously encountered by the Germans – Sixth Guards Army which was to bear the brunt of Hoth's attack held a 60km (37-mile) front. It laid 69,688 anti-tank and 64,340 anti-personnel mines in its first defence line, backed up by a further 20,200 anti-tank and 9097 anti-personnel mines in the second defence line. Mine-laying was concentrated along the anticipated German lines of advance where mine densities averaged between 1400 and 2000 anti-tank mines per kilometre (0.6 mile) of front.

Minefields played an important role in the Soviet defences in the south of the salient, where the defenders laid a total of 292,000 anti-tank mines and 306,000 anti-personnel mines in the sector held by the Voronezh Front. During the battle Soviet engineers laid more than 55,000 additional mines, primarily by mobile obstacle detachments operating under the command of 5th Engineer Sapper Brigade and 42nd Engineer Brigade, in addition to the detachments formed by divisional engineer units. An estimated 113 German tanks, 30 assault guns and 73 other vehicles were destroyed in the minefields laid by mobile obstacle detachments during the battle. According to Soviet sources, the Germans lost a total of about 630 AFVs in all the Soviet minefields on the south face of the salient during the period 5–17 July.

Mines were a particularly important weapon against the Tigers whose armour made them invulnerable to most of the Red Army's tank and anti-tank guns, except at very close range. Minefields were also responsible for inflicting a significant number of casualties – Soviet sources claimed that these were the equivalent of two infantry battalions (approximately 450 men) on the first day of the offensive on the south face of the salient. (The toll included a senior officer – Lieutenant General Schäfer, commander of the 332nd Infantry Division of XLVIII Panzer Corps.)

In addition to conventional mines, the defenders made extensive use of artillery shells – these were fitted with pressure fuses and buried nose-up. The heaviest of these, such as the 43.5kg (96lb) 152mm (6in) and 98.7kg (217lb) 203mm (7.99in) shells were capable of destroying rather than simply disabling most German AFVs. (Fortieth Army alone used 6377 shells to supplement its 130,000 anti-tank and anti-personnel mines.) Other unconventional devices included incendiary mines of up to 20 petrol-filled bottles surrounding an anti-personnel mine which were ignited when the mine detonated.

ARTILLERY

As always, the Red Army gave a high priority to deploying artillery in defence of the salient and had brought in a total of 31,000 guns and mortars by the time that the German offensive began. Whilst the heaviest concentrations were in the north of the salient, Sixth Guards Army in the south deployed 1682 guns and mortars – 316 guns and howitzers (122mm/4.8in and 152mm/6in), 573 anti-tank guns of all calibres and 793 mortars (82mm/3.22in and 120mm/4.72in). These were supported by 88 'Katyusha' salvo rocket launchers.

The anti-tank guns were a key element in the defence lines – many were deployed in mutually supporting anti-tank strongpoints (ATSPs) designed for all-round defence. These were supported by infantry detachments and sappers with anti-tank mines. Each ATSP contained:

- Four to six anti-tank guns (usually 45mm/1.8in and 76.2mm/3in)
- Six to nine anti-rifles
- Two to three medium or heavy machine-guns
- Three or four light machine-guns

All Soviet artillery was integrated into the overall anti-tank defence plan and was positioned to provide the maximum possible fire support for the anti-tank strongpoints. In the event of a German breakthrough, all 122mm (4.8in) and 152mm (6in) guns and howitzers in the threatened sector were to be used in the direct-fire anti-tank role. (Although the majority of these were low-velocity weapons that were too inaccurate to make ideal anti-tank weapons, their heavy shells were effective against most German AFVs at close range.) In similar fashion, all AA guns were earmarked to operate in the anti-tank role if necessary.

The most powerful artillery concentrations are useless without adequate ammunition, and stockpiling sufficient supplies to meet the anticipated demand was a massive task which placed a heavy strain on the limited rail network serving the salient. In this respect, as in many other aspects of defence preparations, the repeated delays in launching the German offensive were of immense value to the Red Army.

ARMOUR

Stavka appreciated that the German offensive could not be defeated solely by even the most formidable fixed defences and ensured that substantial armoured forces totalling almost 3500 AFVs were deployed within the salient. Some were positioned in direct support of the defence lines, but a high proportion of the available armour was concentrated in the First and Second Tank Armies. These formations were intended to defeat the main

Clearly visible on this Tiger I is the *Zimmerit* anti-magnetic mine paste. *Zimmerit* acted as a barrier between a magnetic mine and the tank's armour. Soviet infantry tank-hunting teams used mines to great effect.

German thrusts, leaving the 1600-plus AFVs of the Steppe Front (including Fifth Guards Tank Army) to spearhead the subsequent counter-offensives.

THE GERMAN PLAN

In accordance with the basic principles laid down by von Manstein, Colonel General Hoth was determined to ensure that Fourth Panzer Army would use its exceptional armoured strength to achieve a rapid breakthrough of the formidable Soviet defences. This aim was passed down to corps and divisional level, and *Das Reich's* staff now set about the detailed planning required to put it into practice.

The massive Soviet defensive belts represented the greatest challenge – it was clearly going to be difficult to break through such elaborate field defences at all. To do so without losing the momentum of the attack was incredibly demanding, even for an elite formation such as *Das Reich*. There were no specialized assault tanks or mine-clearing AFVs such as the AVREs and Sherman Crabs employed so effectively by 79th Armoured Division in the Normandy campaign. (Model's Ninth Army attacking the northern sector of the Kursk salient did have three companies of the complex and unreliable Borgward B.IV demolition charge laying vehicles. These were intended to move up to the edge of minefields under radio control and drop their demolition charges to detonate nearby mines, creating a gap for AFVs and infantry.)

In the absence of specialized armour, it was down to the division's over-worked *pionere* (assault engineers) to 'feel' their way through each minefield, cautiously probing for mines with bayonets or two thin steel rods 1m (40in) long. Each mine had to be carefully lifted and disarmed by hand before the cleared lanes were marked, usually with white tape. (On occasions there was no time for this and the *pionere* themselves would lay down to act as markers.)

Known Soviet bunkers and strongpoints were targeted by the preliminary artillery bombardment, but it was recognized that this was likely to be of limited effectiveness, especially against well-constructed bunkers with good overhead cover. Such targets would often have to be destroyed by *pionere* assault teams with flamethrowers and demolition charges, although direct fire from heavy infantry guns could be devastating, particularly when using the muzzle-loaded 90kg (198lb) Stielgranate 42 stick bomb.

Under the final divisional plan, *Das Reich* was to operate in distinct tactical elements: assault, exploitation, reserve and main body. The assault element consisted of *Deutschland*, an engineer company, several batteries of assault guns and the divisional Tiger company. The panzer regiment, the armoured panzergrenadier battalion, the self-propelled artillery and an engineer company composed the exploitation group. *Der Führer*, the panzerjäger battalion, the reconnaissance battalion, and the rest of the engineer battalion constituted the tactical reserve while all remaining artillery and support elements were grouped in the division's main body.

Operation 'Citadel'

Although Hitler's Operational Order No. 6 had stressed the need for the swift elimination of the Kursk salient, the offensive was repeatedly postponed to the growing alarm of many German commanders.

Hitler's motive for ordering these repeated delays was primarily a belief that the new Panthers and Ferdinands just entering service were essential for the success of the offensive. His belief that they could be rushed into service to ensure a decisive breakthrough of the ever-improving Soviet defences prompted an outburst from Guderian: 'I don't regard the new Panther or Ferdinand as ready for active service. They are still suffering from numerous teething troubles as is perfectly natural with such new types – and we can't possibly clear these up in five or six weeks!'

Despite the protests from Guderian and other senior officers, Hitler was adamant that the offensive would not be launched until both new types were available in quantity, thus giving the Soviets more invaluable time to strengthen their defences. (Ironically, the 200 Panthers and 90 Ferdinands committed to the operation proved to be more trouble than they were worth, suffering appalling mechanical problems and depressingly frequent breakdowns. All too many broken down vehicles had to be abandoned due to the difficulties in recovering these heavy AFVs with equipment designed for the much lighter Panzer IV and StuGs.)

DAS REICH – RE-EQUIPMENT DELAYS

In March 1943, the 1st Battalion of the division's Panzer Regiment had been transferred to Germany to convert from the Panzer III to the new Panther, but this proved to be an unexpectedly lengthy process. This was due to two main factors – the need to absorb and train replacements for the casualties sustained in Russia, and delays in Panther production as the worst of the type's teething problems were rectified. As a result, the unit did not complete its conversion to the Panther until mid-August and only was unable to go into action until 22 August, over a month after the end of Operation 'Citadel'.

INTO ACTION

The much delayed German offensive finally opened on 5 July – *Generaloberst* Model's Ninth Army attacked the northern face of the salient, whilst Hoth's Fourth Panzer Army and Army Detachment *Kempf* assaulted the first Soviet defence line in the south.

Despite the support of all the serviceable Ferdinands, Model's attacks quickly bogged down. In part this was due to the sheer strength of the defences in this sector, but Model's tactics were also partly responsible. He had decided to use his infantry to assault the Soviet lines, with the panzers following up to exploit the breakthrough. This

RIGHT: Although *Das Reich's* casualties were significant, they were not unduly high for a formation that had been involved in heavy fighting for this sort of period. It is worth comparing them with those for the following Allied divisions in the Normandy campaign: 3rd Canadian Division, 2831 (6–12 June 1944); 15th (Scottish) Division, 2720 (Operation 'Epsom', 26–30 June 1944); US 83rd Infantry Division, 5000 (2–12 July 1944).

prevented the attacks from developing any real impetus, reducing them to a process of slow, grinding attrition reminiscent of World War I. The painfully slow German advance was finally halted on 10 July after achieving a maximum penetration of no more than 17km (10.5 miles).

Hoth had adopted very different tactics for the attacks in the south – these were to be carried out by massed panzer formations acting as armoured battering rams, with the infantry following on to hold the newly won ground. His Fourth Panzer Army was able to exploit the shock of these massed armoured assaults and *Luftwaffe* air superiority to achieve a far deeper penetration of the Soviet defences than had been possible on the north face of the salient.

The operational level plan was that west of Belgorod the two panzer corps of Fourth Panzer Army would attack northwards towards Oboyan before swinging to the northeast to intercept and defeat the Soviet armoured reserves around Prokhorovka and then resuming the drive on Kursk. These two corps, XLVIII and II SS Panzer Corps, fielded a total of two panzer, four panzer grenadier divisions and two infantry divisions. Similarly, the III Panzer Corps with its three panzer

II SS-Panzer Corps: Casualties, 5–20 July

DATE	KILLED	WOUNDED	MISSING	TOTAL
5 July	67	223	0	290
6 July	43	180	2	225
7 July	18	103	1	122
8 July	50	186	0	236
9 July	22	127	2	151
10 July	16	94	2	112
11 July	29	181	1	211
12 July	41	190	12	243
13 July	17	44	0	61
14 July	58	229	0	287
15 July	26	88	0	114
16 July	58	166	0	224
17 July	10	23	3	36
18 July	1	8	0	9
19 July	0	2	0	2
20 July	0	5	0	5
Total	456	1849	23	2328

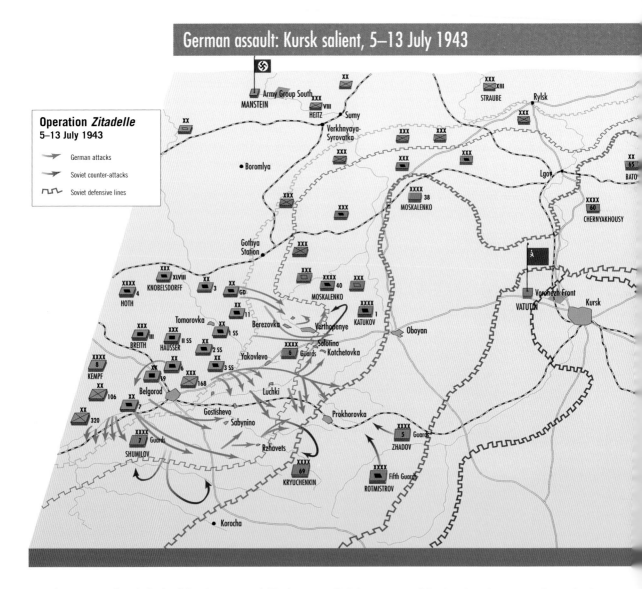

German assault: Kursk salient, 5–13 July 1943

**Operation *Zitadelle*
5–13 July 1943**

→ German attacks

→ Soviet counter-attacks

⌐⌐ Soviet defensive lines

ABOVE: The German attacks on both sides of the salient were crippled by the lack of good infantry formations to support the panzers. Even during the spectacular victories of 1941–42, the German infantry had sustained huge casualties – as early as May 1942, almost all units in the Soviet Union were under strength, reporting a total shortage of 635,000 men. By mid-1943, the situation had worsened, forcing the use of panzer and panzergrenadier units in roles that should have been carried out by infantry. (The problem had even more widespread effects because the *Luftwaffe* was frequently called upon to provide emergency assistance for the hard-pressed infantry instead of attacking key targets behind the Soviet front line.) During the Kursk offensive, German commanders were repeatedly forced to divert armoured units to support weak infantry formations. The effect of this was most serious in the southern sector, where it may have been instrumental in preventing a decisive breakthrough by Fourth Panzer Army.

divisions would lead the attack of Army Detachment Kempf southwest of Belgorod. Three all-infantry corps were assigned to defend the flanks of the penetrations – LII Corps was to cover the western flank of Fourth Panzer Army, whilst XI and XLII Corps protected the flanks of Army Detachment *Kempf*.

Army Group South had 135 Tiger tanks – the equivalent of 3 battalions. During the attack on the south face of the salient these were not employed in mass, as General Guderian had recommended, or even by battalions. Instead, they were employed as individual companies in support of the panzer divisions. This tactical

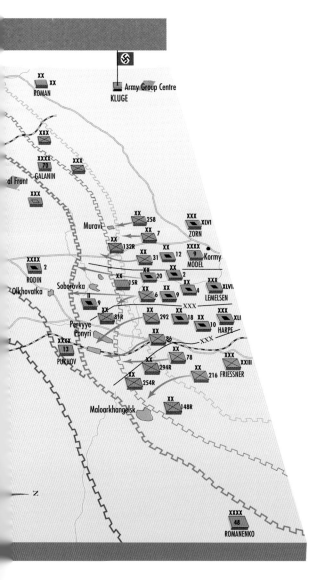

battalion strength with about 50 tanks deployed on a frontage of about 1200m (3937ft). The first wave was formed by two lead companies, preferably including a Tiger company at the point of the arrowhead. Each lead company deployed two platoons of five tanks in a 500m (1640ft) line with the tanks spaced at roughly 50m (164ft) intervals. The two remaining platoons and the HQ section followed in column.

FIRST WAVE

The primary function of the first wave was to suppress the enemy anti-tank defences. The Tigers were especially effective in this role as their 88mm (3.5in) guns significantly out-ranged the Soviet tanks and 76.2mm (3in) anti-tank guns. The second wave included a tank company to provide covering fire for the first wave before attacking the enemy positions together with one or two companies of the division's panzer grenadiers in SdKfz 251 halftracks. The panzer grenadiers would dismount as close as possible to their objectives before dismounting for the final attack covered by fire from assault guns. A third wave, with the remaining tank company and the bulk of the panzer grenadiers, would eliminate any remaining resistance. The flanks were protected by anti-tank guns, operating by platoons and moving by bounds to cover each other's advance.

These were essentially standard German tank-infantry tactics, except for the integration of Tigers in the first wave. Once the attack broke into the Soviet frontline company strongpoints, the tactical drill was to eliminate the defensive positions one by one. Each strongpoint was an essential link in the chain of interlocking fire, dependent on its neighbours to lay down a curtain of fire across its front. Thus, immediately after capturing each strong point, the Germans intended to widen the breach by rolling up the neighbouring strong points from each flank. The Soviets were well aware of the German tactics and attempted to forestall them by:
■ Preparing deep defensive belts rather than thin defence lines
■ Providing defence in depth totalling eight defence belts
■ Deploying massed artillery and anti-tank guns
■ Holding reserves in readiness for counter-attacks
■ Laying extensive minefields to hinder the German ability to manoeuvre.

employment was favoured by the TOE (Table of Operation and Equipment); each of the SS divisions and Panzergrenadier Division *Grossdeutschland* had a company of Tigers. (The independent Heavy Panzer Battalion 503 was used by III Panzer Corps in the same way to support its panzer divisions.) However, the 200 Panthers of Panzer Battalions 51 and 52 (Panzer Regiment 39) were employed en masse in support of Panzergrenadier Division *Grossdeutschland*.

Whenever the terrain and the Soviet minefields on the south face permitted, the Germans attacked in *Panzerkeil*, or arrow, formation with tanks in the lead. The panzers usually advanced in roughly

Das Reich's Actions on the 'Road to Prokhorovka'

From the first hours of the offensive, *Das Reich* played a key role in breaking through the Soviet defences. Although the panzers were used wherever possible to minimize infantry casualties and speed up the advance, there were times when there was no alternative but to rely on the over-worked panzergrenadiers.

On the night of 4/5 July, pioneers and the panzergrenadiers of *Deutschland*'s 2nd and 3rd Battalions infiltrated the Soviet outpost line before 12 flamethrower teams led the attack on the forward defences. Hans Huber was a member of one of those teams and later recalled just how effective they had been: 'The enemy artillery fire forced us to take cover. One might almost add "thank God" because the equipment was damnably heavy. ... Section commander Kessler grew impatient. He ordered me to get the flamethrower ready and we worked our way forward into the trenches ahead of us. I fired a burst of flame as we approached every zig-zag in the trench and at every enemy strongpoint. It was a strange feeling to serve this destructive weapon and it was terrifying to see the flames eat their way forward and envelop the Russian defenders. Soon I was coloured black from head to foot from the fuel oil and my face was burnt from the flames which bounced back off the trench walls. ... I could hardly see. The enemy could not fight against flamethrowers and we made good progress taking many prisoners.'

After the cumulative effect of hundreds of such actions had broken through the first Soviet defensive belt, the panzer regiment took the lead against the T-34s and US-supplied M3s Lee tanks of Katukov's First Tank Army which counter-attacked in an attempt to halt the German advance. A war correspondent with *Das Reich*'s Tiger company wrote of a typical action fought during the afternoon: 'we see two T-34s stationed on the

An armoured column of Panzer IVs and some lighter vehicles belonging to *Das Reich* prepares to move off, sometime during the Kursk offensive. On the left is a Panzer VI Ausf E Tiger tank.

dominating heights to the north. Their first shells land near us. Then a number hit our front plate and hull, but they do not do us much harm. We move forward a little to take up a better firing position. … A hit! We pursue. The first T-34 is burning. Our neighbour has destroyed the second, which has also started to burn. After we have moved forward another 500 metres about 40 enemy tanks appear on the horizon. They advance past the blazing wrecks of the first two we have destroyed, then stop, shoot, then move forward again, this time firing quickly on the move, one shell after another. Again rounds splatter against our front plate and hull but do us no harm.'

The success of the first day's attacks owed much to the *Luftwaffe*'s highly effective close air support, especially the Fw 190F fighter bombers of *Schlachtgeschwader* 1, which made repeated attacks on Soviet artillery and anti-tank positions with SD1 and SD2 cluster bombs. These fragmentation weapons inflicted heavy casualties on the Soviet gun crews because, although they occupied well dug-in and carefully camouflaged positions, the majority of these lacked overhead protection.

STAVKA'S RESPONSE

Although Fourth Panzer Army's advance was slowed by the extensive Soviet defences, its progress during the first few days of the offensive caused considerable alarm at the HQ of General Vatutin's Voronezh Front, which was responsible for the defence of the southern sector of the salient. On the evening of 6 July, Vatutin sent a report on the situation to Stalin, with a request for reinforcements totalling four tank corps and substantial additional air support. Marshal Vasilevsky, *Stavka*'s representative at the Front HQ, strongly backed this assessment and actually went further, recommending that two additional tank corps should be sent if the Front was to 'conduct further active operations.'

Stalin accepted their recommendations and ordered the transfer of Fifth Guards Tank Army from General Konev's Steppe Front to the Voronezh Front. (The Army was also substantially reinforced by the addition of XVIII Tank Corps to its original XXIX Tank Corps and V Guards Mechanized Corps.) As Hoth had foreseen, *Stavka*'s intention was to concentrate Rotmistrov's command around Prokhorovka and attack Fourth Panzer Army's right flank as it attempted to force its way across the River Psel. Konev was unhappy at what he feared was the first of a series of moves that would gradually whittle away the strategic reserves and threaten the success of the offensives which were

Destroyed Tanks in Army Group South, 5–17 July 1943

UNIT	FLAMM.PZ	PANZER III	PANZER IV	PANTHER	TIGER I	STUG
3rd Panzer Division	–	6	3	–	–	–
11th Panzer Division	–	2	3	–	–	–
Grossdeutschland	–	3	16	–	–	1
LAH	–	1	9	–	1	1
Das Reich	–	1	6	–	1	1
Totenkopf	–	2	8	–	1	1
6th Panzer Division	3	9	13	–	–	–
7th Panzer Division	–	8	2	–	–	–
19th Panzer Division	–	8	19	–	–	–
503rd Panzer Abteilung	–	–	–	–	3	–
10th Panzer Brigade	–	–	–	44	–	–
911th StuG Abteilung	–	–	–	–	–	3
905th StuG Abteilung	–	–	–	–	–	5
228th StuG Abteilung	–	–	–	–	–	1
393rd StuG Bttr	–	–	–	–	–	5

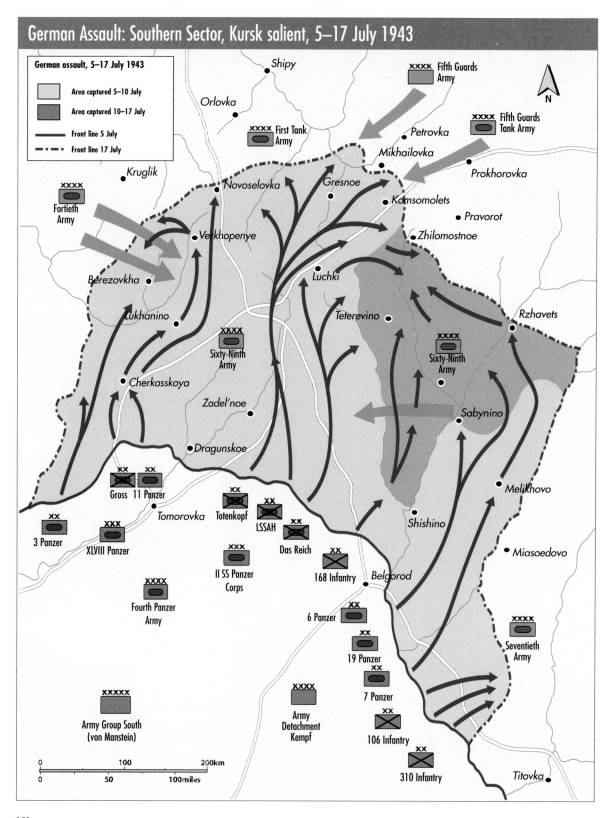

German Assault: Southern Sector, Kursk salient, 5–17 July 1943

German assault, 5–17 July 1943

Area captured 5–10 July

Area captured 10–17 July

Front line 5 July

Front line 17 July

Shipy

Orlovka

Kruglik

Fortieth Army

First Tank Army

Novoselovka

Gresnoe

Petrovka

Mikhailovka

Prokhorovka

Fifth Guards Army

Fifth Guards Tank Army

Komsomolets

Pravorot

Verkhopenye

Zhilomostnoe

Berezovkha

Luchki

Rzhavets

Lukhanino

Teterevino

Sixty-Ninth Army

Sixty-Ninth Army

Cherkasskoya

Zadel'noe

Sabynino

Dragunskoe

Melikhovo

Gross 11 Panzer

Tomorovka

Totenkopf

LSSAH

Shishino

Miasoedovo

3 Panzer

XLVIII Panzer

Das Reich

168 Infantry

Belgorod

II SS Panzer Corps

Fourth Panzer Army

6 Panzer

Seventieth Army

19 Panzer

Army Detachment Kempf

7 Panzer

106 Infantry

Army Group South (von Manstein)

310 Infantry

Titovka

0   100   200km

0   50   100miles

scheduled to be launched after the German attack had been defeated. Konev argued his case so forcefully that Stalin phoned him with an abrupt order to get Fifth Guards Tank Army moving immediately.

After speaking to Konev, Stalin contacted Rotmistrov to discuss the details of the redeployment. He agreed to Rotmistrov's proposals to move the entire force by road in a series of forced marches with air cover being provided from dawn to dusk. The move was made along three routes in a 30–35km (19–22 mile) sector – even spread over two or three routes, each tank corps occupied 20–30km (12–19 miles) of road.

Even if the redeployment could be carried out with textbook precision, Fifth Guards Tank Army could not reach the front line for several days and the Red Army formations defending the south of the salient were in desperate need of more immediate armoured support. *Stavka* therefore ordered that II Tank Corps should be released from the Southwestern Front, whilst Voronezh Front was to commit II Guards Tanks Corps from its reserves. (Both formations would be attached to Fifth Guards Tank Army at Prokhorovka.)

DAS REICH VERSUS TWO SOVIET TANK CORPS

II Guards Tank Corps had already been moved to the area south of Prokhorovka during the night of 5/6 July in preparation for attacks against the right flank of II SS Panzer Corps. During the following afternoon, it counter-attacked *Das Reich* east of Luchki in conjunction with 96th Tank Brigade, temporarily halting the German advance on the west bank of the Lipovyi Donets River.

OPPOSITE: Von Manstein's decision to commit his panzers en masse from the beginning of the offensive allowed his forces to make good progress through the Soviet defences in the south of the Kursk salient. His initial attacks also benefitted greatly from the *Luftwaffe*'s highly effective close air support. However, the chronic shortage of German infantry meant that armoured divisions were constantly being held back to protect the flanks of the penetration, which in turn fatally slowed the rate of advance. However, the sheer tactical skill of elite formations such as II SS Panzer Corps very nearly succeeded in overcoming even these crippling problems – by 10/11 July, the Soviet position was so critical that Stalin was demanding hourly situation reports. Even in the immediate aftermath of the fighting at Prokhorovka, he was so concerned at the possibility of a German breakthrough that he appointed Zhukov as *Stavka* representative to oversee future operations in that sector.

7 July – Additional heavy tank battles took place between Kalinin and Teterevino, including an attack at 10:30 by II Tank Corps. *Das Reich* successfully beat off these attacks and by the end of the day had advanced over 10km (6.2 miles), reaching Teterevino on the road to Prokhorovka. This represented a significant gain after a hard day's fighting, but it was still 10km (6.2 miles) short of the third defence line, the objective for the day.

8 July – During the morning, II Tank Corps moved into position south of Prokhorovka just behind the third defence line before launching a series of attacks against the infantry units of *Das Reich* south of Teterevino. Throughout the afternoon it was involved in further attacks at Teterevino, North Luchki and Kalinin. In one of these attacks, 20 Soviet tanks broke through and dispersed the divisional artillery. The German infantry received unexpected help from *Totenkopf*'s reconnaissance battalion, which happened to be passing through on the way north. To meet the simultaneous heavy attacks from the northwest, the northeast, and the east, II SS Panzer Corps had to use all its reserves. In a typical emergency *Das Reich* formed an ad hoc battle group comprising its engineer battalion, 627th Engineer Battalion, and 818th Artillery Regiment's 3rd Battalion to plug a dangerous gap opening up between it and *Totenkopf*, which was still defending the corps' east flank. The division was able to hold its positions only thanks to superb *Luftwaffe* close air support which broke up several threatening tank attacks.

Despite the constant Soviet attacks, the division was able to carry out a classic infiltration operation that evening. Entirely unofficially, *Der Führer's* commander, *Sturmbannfuhrer* Stadler, had established a regimental intelligence cell which included six Russian *Hiwis* (volunteers) who listened in to Red Army radio traffic. This small unit had a fine track record and it proved to be invaluable on the afternoon of the 8th, when it picked up a furious exchange of messages from the Soviet units opposing them going all the way up to corps and army level.

Everyone was blaming each other for the failure to halt the German advance and making accusations that promised reinforcements had been 'stolen' by other commanders. Stadler recognized that the disorganization and panic in the Soviet command structure was ripe for exploitation and detailed the 1st Battalion's No. 3 Company under *Hauptsturmführer* Lex to carry out a night raid. The battered Red Army units holding this sector of the front were far from alert and the raiding force was able to slip through their positions undetected before launching an attack that captured a Soviet brigade HQ complete with the brigadier, his staff and the HQ defence detachment.

9 July – On the corps' east flank, *Das Reich*, with one regiment of 167th Infantry Division

attached, spent the day fending off tank-supported attacks from the north and northeast by elements of V Guards Mechanized Corps and II Tank Corps. Some of the attacks from the direction of Prokorovka were substantial affairs, including up to 100 tanks. Further south other elements of II Tank Corps attacked two regiments of 167th Infantry Division defending Fourth Panzer Army's east flank.

11 July – Two regiments of *Das Reich* attacked II Tank Corps south of Prokhorovka at 09:30. The

SS-Brigadeführer Sylvester Stadler (30 December 1910 – 23 August 1995)

Sylvester Stadler was born in Fohnsdorf, Austria. After training as an electrician, he joined the Austrian SS at the age in 1933 and was posted to Germany for training before becoming a platoon commander in *SS-Standarte 'Deutschland'*. In 1935, Stadler was selected to attend the *SS Junkerschule* at Bad Tölz, from which he graduated with the rank of *SS-Untersturmführer*. During the Polish campaign he served in Panzer Division *Kempf* (a mixed formation of SS and Army panzer and infantry units).

In 1940 he was posted to the *Der Führer* regiment as a company commander. He was transferred back to *Deutschland* on 20 September 1941 as the commander of its second battalion. Whilst in this post, Stadler was wounded for the first time and evacuated to Germany to recover. During his convalescence, he served as an instructor at the *SS Junkerschulen* at Bad Tölz and Braunschweig. Stadler returned to *Der Führer* in March 1942, becoming commander of its second battalion. He led the battalion during the Fourth Battle of Kharkov in February/March 1943 and was awarded the Knight's Cross for his outstanding leadership in June of that year. In the same month, he took over as regimental commander of *Der Führer*, leading it throughout Operation 'Citadel' and the subsequent defence of the river Mius, for which he was awarded the Oak Leaves to his Knight's Cross in September 1943.

In July 1944, Stadler took command of 9. SS-Panzer-Division *Hohenstaufen*, becoming one of the youngest German divisional commanders. He was severely wounded during the Normandy campaign, but after recovering from his wounds, he returned to command *Hohenstaufen* from October 1944 until the end of the war, reaching the rank of *SS-Brigadeführer*. He was awarded the Swords to the Knight's Cross on 6 May 1945, two days before he surrendered to US forces in Austria.

struggle lasted more than three hours, as the panzers struggled to deal with the supporting Soviet anti-tank guns. Finally the Soviets were forced to withdraw northeast toward Prokhorovka. Further to

the south at Ivanovka other elements of *Das Reich* had to fight off several attacks by II Tank Corps and were able to advance only about 1km (0.6 mile) to the east.

Prelude to Prokhorovka

Rotmistrov and his HQ staff arrived at Prokhorovka in the early hours of 11 July, well ahead of the main body of Fifth Guards Tank Army to receive Vatutin's orders that by 10:00 the next morning he was to 'deliver a counterstroke in the direction of Komsomolets State Farm and Pokrovka and, in conjunction with Fifth Guards Army and First Tank Army, destroy the enemy in the Kochetovka, Pokrovka and Greznoye regions and do not permit him to withdraw in a southerly direction.'

However, these orders were rapidly overtaken by events as II SS Panzer Corps had managed to continue its advance on 11 July and actually occupied the ground to the west and southwest of Prokhorovka, which Rotmistrov had earmarked as the jumping-off areas for his attacks. There was a potentially disastrous breakdown in the Soviet command and control system at this stage; it seems that no-one informed Voronezh Front HQ or Fifth Guards Tank Army of the rapidly changing situation. Rotmistrov had been informed that II Tank Corps and II Guards Tank Corps would be under his command and intended to deploy them in his first echelon, alongside XVIII and XXIX Tank Corps. V Guards Mechanized Corps would form the second echelon, leaving only a small task force under his deputy, Major-General Trufanov, to act as a reserve.

Having completed his deployment plans, Rotmistrov was visited at about 19:00 by Marshal Vasilevsky, who was keen to inspect the areas from which Fifth Guards Tank Army would attack. They set off with a small escort in a convoy of jeeps. On arrival at the selected positions, Vasilevsky was furious to see what he thought was Fifth Guards Tank Army milling around in the open. However, Rotmistrov quickly identified the tanks as German and the two commanders beat a hasty retreat.

The unexpected German advance meant that Rotmistrov had to hurriedly revise his plans for the

following day. As he wrote: 'we had to prepare for the offensive anew; in particular select artillery firing positions and deployment and attack lines. In the compressed time, we had to refine missions, organize cooperation between corps and units, revise the schedule for artillery support, and do all to facilitate the precise command and control of forces in combat.'

COMFORTABLE ASSUMPTIONS

It seems that both Rotmistrov and Vasilevsky had assumed that the German advance towards Prokhorovka had been virtually halted and were working on the basis that Fifth Guards Tank Army would be the spearhead of a counter-offensive to at least retake the territory lost since 5 July. These comfortable assumptions were abruptly shattered by their uncomfortably close encounter with *Leibstandarte* on the evening of 11 July. Suddenly, it was apparent that II SS Panzer Corps was still a force to be reckoned with and was still quite capable of taking Prokhorovka.

This realization prompted Rotmistrov to issue tactical instructions based on earlier discussions with Vatutin and Vasilevsky at Voronezh Front's HQ. He recalled Vasilevsky's question: 'The German tank divisions possess new heavy Tiger tanks and Ferdinand self-propelled guns. Katukov's tank army has suffered considerably from them. Do you know anything about this equipment and how do you feel about fighting them?

'We know, Comrade Marshal. We have received tactical and technical information about them from the Steppe Front staff. We have also thought about means for combating them. … The fact is that the Tigers and Ferdinands not only have strong frontal armour, but also a powerful 88mm gun. … In that regard they are superior to our tanks which are armed with 76mm guns. Successful struggle with them is possible only in circumstances of close combat, with exploitation of the T-34's greater manoeuvrability and by flanking fire against the side armour of the heavy German machines.'

Musing on this, Vatutin had commented: 'In other words, engage in hand-to-hand fighting and board them.'

Private, Panzer crew

Initially, the only items of **Waffen-SS** *camouflage clothing were helmet covers and smocks. However, their popularity was such that the range was soon expanded to include overalls for AFV crews, which were made in a variety of camouflage patterns. This private wears one of the predominantly brown 'autumn' patterned sets of overalls, without any personal equipment to give maximum freedom of movement in the cramped confines of the tank's fighting compartment.*

Ironically, much of this discussion was based on the false premise that, as an elite formation, II SS Panzer Corps must be equipped with large numbers of Tigers and Ferdinands. The reality was very different – even at the beginning of Operation 'Citadel', the corps fielded a total of just 57 Tigers and not a single Ferdinand. (All 90 of the latter were assigned to support Ninth Army's assault on the north of the salient.) It is likely that the *Schurzen* – skirting armour – fitted to the majority of the German tanks and assault guns was partly responsible for the confusion, but there is an equally strong possibility that Soviet commanders were keen to report masses of Tigers and Ferdinands to justify their own heavy losses and inability to halt the German advance.

Whilst not going so far as Vatutin's comment had implied, the orders sent out to all Romistrov's units emphasized the need to advance swiftly against the German armour and close the range as quickly as possible so that massed fire from the T-34s and T-70s could 'swamp' the panzers. If they could achieve surprise, most would have a reasonable chance of surviving for the five minutes or so that it would take to get into close combat. In order to give his units the best chance of achieving that surprise, Rotmistrov brought forward his attack to 06:30.

PINCER THRUST

However, II SS Panzer Corps was not his only problem: away to the south, III Panzer Corps spearheading the advance of Army Detachment *Kempf*, was finally making real progress towards Prokhorovka. During the night of 11/12 July III Panzer Corps seized the bridge over the Donets at Rzhavets and pushed north to within 20km (12.4 miles) of Prokhorovka, endangering Fifth Guards Tank Army's left flank. Rotmistrov now had to face the strong possibility that he would have to fight both panzer corps.

To counter this threat, Major General Trufanov was assigned V Guards Mechanized Corps (less 10th Guards Mechanized Brigade and 24th Guards Tank Brigade), together with most of the smaller units from the Army's second echelon. This was a well balanced force with about 120 AFVs, which made a powerful reinforcement for the 10 anti-tank regiments fielding a total of at least 200 guns (a mixture of 45mm/1.8in and 76.2mm/3in weapons), which were being deployed in the attempt to halt III Panzer Corps.

THE BATTLEFIELD

For both sides the best terrain for the attack was a narrow strip of land southwest of Prokhorovka; 6–8km (3.7–4.9 miles) wide, it formed a corridor of good tank country. (Its importance had been recognized before the German offensive began and a huge anti-tank ditch forming part of the third line of Soviet defences ran south-east from the Oktabrisky State Farm, cutting across the southern half of the corridor.) The area was bounded on the north by the River Psel and to the south by the embankment of the Belgorod to Kursk railway line, which formed a significant obstacle for tanks. The area was generally flat and largely covered with standing crops of wheat and maize, broken by gullies where the ground sloped down to the river and the villages of Mikhailovka, Prelestnoye and Petrovka, which straggled along its southern bank. A further village, Polyzhaev, lay on the northern bank – all four settlements had been fortified and were occupied by infantry, artillery and Katyusha rocket batteries.

The terrain north of the Psel was also relatively good going for tanks, but the river itself was a serious obstacle, especially in its swollen state after the rain of the last few days. However, *Totenkopf* had seized a small bridgehead on the far bank on 10 July and had completed one bridge capable of carrying medium tanks by the following day. South of the railway embankment, the terrain was less favourable for armoured units, with scattered woodland, marshes and small ravines.

At dawn on 12 July, both sides were occupying less than ideal positions. II SS Panzer Corps certainly had its problems – on the left, *Totenkopf* was still penned into a small bridgehead across the Psel, whilst in the centre *Leibstandarte*'s advance of the previous evening had left it in a narrow salient west of Prokhorovka. Its left flank was uncomfortably exposed to fire and counter-attacks from the fortified villages along the banks of the Psel. *Das Reich* was unable to add its weight to the advance on Prokhorovka because it was committed to holding the corps' right flank against repeated

A heavily camouflaged Das Reich *tank receives the final touches from its crew during the Kursk campaign.*

Soviet attacks. (Its positions ran south from the area of Storozhevoe, roughly parallel to the Belgorod–Kursk highway.)

The Red Army's positions were equally problematic – in the north, all attempts to crush the German bridgehead on the east bank of the Psel had failed and there was a real risk that *Totenkopf* might break through the defences of the badly battered 52nd and 95th Guards Rifle Divisions to threaten Fifth Guards Tank Army's right flank. In the centre, the forward positions of *Leibstandarte* were little more than 3km (1.8 miles) from Prokhorovka itself, confining the bulk of Rotmistrov's forces to cramped deployment areas to the northwest and north of the town. A German breakthrough in this sector would open up the possibility of a swift advance along the Belgorod-Kursk highway. By taking this route, II SS Panzer Corps would have to breach only a single further defence line before reaching Kursk. Rotmistrov's left flank was held by the depleted II Guards Tank Corps and II Tank Corps. They were in the unenviable situation of trying to attack *Das Reich* whilst their left flank was under increasing threat from the advance of III Panzer Corps, which was now barely 20km (12.4 miles) south of Prokhorovka.

THE CORRELATION OF FORCES

For many years after the war, the accepted wisdom was that Prokhorovka was a truly epic tank battle with huge forces deployed on each side. Soviet sources claimed that the total German strength was in the order of 600 AFVs, including 100 Tigers and Ferdinands. The reality was very different – by 12 July, II SS Panzer Corps deployed no more than 294 operational tanks and assault guns. The probable breakdown for this was:

- *Totenkopf* – 121, including 10 Tigers
- *Leibstandarte* – 70, including 4 Tigers
- *Das Reich* – 103, including 1 Tiger.

Rotmistrov's forces probably fielded a total of approximately 850 AFVs, including: 501 T-34s, 261 T-70s, 31 Churchill Mark IVs and 21 KV-1s, plus 10 SU-76s, 11 SU-122s and 11 SU-152s. This total was reduced to perhaps 730 tanks and assault guns by the detachments sent to halt the advance of III Panzer Corps, but still gave Fifth Guards Tank Army an overall numerical superiority of almost 2.5:1.

A camouflage-painted Stug III from the *Das Reich* division prepares to move in support of an infantry attack. The weather during Operation 'Citadel' was mixed, with sharp but heavy showers often creating muddy conditions.

Tank Battle at Prokhorovka: German and Soviet Forces Compared

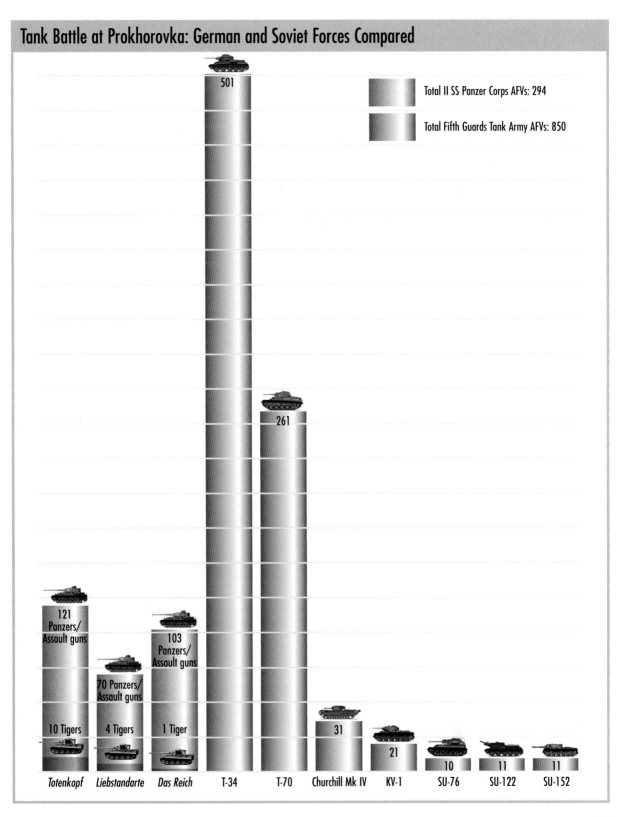

Total II SS Panzer Corps AFVs: 294

Total Fifth Guards Tank Army AFVs: 850

501

261

121 Panzers/ Assault guns

103 Panzers/ Assault guns

70 Panzers/ Assault guns

10 Tigers

4 Tigers

1 Tiger

31

21

10

11

11

Totenkopf *Liebstandarte* *Das Reich* T-34 T-70 Churchill Mk IV KV-1 SU-76 SU-122 SU-152

The Battle: 12 July 1943

Traditionally, the battle has been described in terms of the fighting in the centre where Rotmistrov's XVIII and XXIX Tank Corps, supported by elements of V Guards Mechanized Corps, clashed with *Leibstandarte*.

However, this was only one element in an interrelated series of actions which would have a major impact on the outcome of the German offensive in the southern sector of the salient. In order to give a clearer impression of the scale of the fighting which began on the morning of 12 July, each sector of the battlefield will be examined in turn, starting in the north.

NORTH OF THE PSEL

On the morning of 12 July, *Totenkopf* was holding what was still a very constricted bridgehead no more than 4km (2.5 miles) wide across the Psel. By dawn, engineers had finally completed a Class 60 bridge to allow the division's 10 serviceable Tigers to cross. Except for a small detachment of assault guns and infantry guarding the bridges, the bulk of the division was now concentrated on the north bank ready to renew attacks on the remnants of the 52nd and 95th Guards Rifle Divisions, which were supported by the equally battered 31st Tank Corps.

Both sides recognized the importance of Hill 226, which dominated the terrain in this sector. The hill and the surrounding area were a maze of minefields and field fortifications, but throughout the morning its already weakened garrison was subjected to a series of air attacks, followed by artillery and rocket bombardments. *Totenkopf's* infantry began the assault on the hill with armoured support at 1200, and despite fanatical resistance had taken the position in little more than an hour's fighting. Attempts to follow up this success with an immediate advance north eastwards bogged down in further Soviet defences, whose defenders were strengthened by the arrival of elements of Zhadov's Fifth Guards Army.

Totenkopf regrouped during the late afternoon and called in a major strike by the *Luftwaffe,* which was made with great effect in the early evening despite the rapidly worsening weather, with thunderstorms and heavy rain sweeping across the battlefield. Following up this success, the Tiger company led an all-out attack, and by dusk it had taken the village of Polyzhaev and was poised to cut the Kursk-Belgorod highway.

Rotmistrov was understandably alarmed by this advance, which threatened to outflank his positions. Although Fifth Guards Army was steadily reinforcing the units fighting north of the Psel, it was weak in artillery and had no tanks as its sole armoured formation, X Tank Corps, had been transferred to the command of Voronezh Front on 7 July. Faced with what could rapidly develop into a real catastrophe, Rotmistrov had no choice but to use his own small reserve force and ordered V Guards Mechanized Corps to deploy the remnants of its 24th Tank Brigade around the Voroshilov State Farm astride the Kursk-Belgorod highway with orders to stop *Totenkopf's* advance at all costs.

PROKHOROVKA

The action around Prokhorovka itself began near dawn, when Soviet armour made a reconnaissance in force, probing *Leibstandarte's* positions. This was rapidly driven off without loss and the Germans were able to complete refuelling and loading ammunition in preparation for their own attack, which was scheduled for 06:00. (Having filled their ammunition racks, most experienced tank and assault gun crews stowed extra ammunition in any available space. This proved to be a wise precaution and was one of the factors contributing to the day's very high Soviet losses.)

As these preparations were completed, the Soviet tank crews were also carrying out their final checks. Rotmistrov had decided to set up his HQ at the command post of XXIX Tank Corps, which was

OPPOSITE: Throughout 12 July, the bulk of Rotmistrov's Fifth Guards Tank Army made repeated efforts to break through on the narrow sector of front held by *Leibstandarte,* but was repelled with horrendous losses. *Das Reich* was also subjected to a series of ferocious attacks by II Tank Corps and II Guards Tank Corp,s which prevented it from adding any appreciable weight to *Leibstandarte's* attempts to take Prokhorovka. (On the northern flank, *Totenkopf* made only slow progress against fierce opposition.) By the end of the day, both sides were exhausted and II SS Panzer Corps had effectively been fought to a standstill.

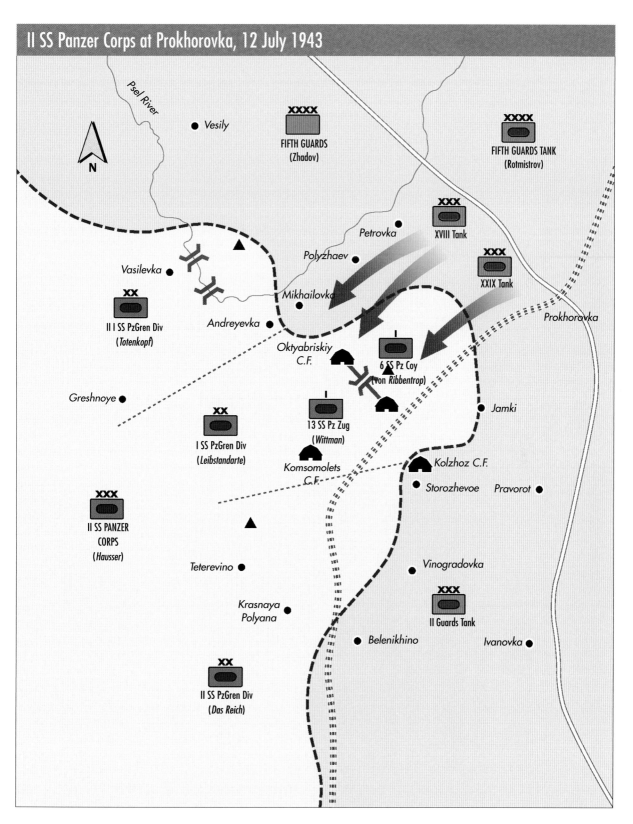

II SS Panzer Corps at Prokhorovka, 12 July 1943

Psel River

Vesily

XXXX
FIFTH GUARDS
(Zhadov)

XXXX
FIFTH GUARDS TANK
(Rotmistrov)

N

Vasilevka

Petrovka

Polyzhaev

XXX
XVIII Tank

XXX
XXIX Tank

XX
II I SS PzGren Div
(*Totenkopf*)

Mikhailovka

Andreyevka

Prokhorovka

Oktyabriskiy
C.F.

I
6 SS Pz Coy
(von *Ribbentrop*)

Greshnoye

XX
I SS PzGren Div
(*Leibstandarte*)

I
13 SS Pz Zug
(*Wittman*)

Jamki

Komsomolets
C.F.

Kolzhoz C.F.

XXX
II SS PANZER
CORPS
(*Hausser*)

Storozhevoe

Pravorot

Teterevino

Vinogradovka

Krasnaya
Polyana

XXX
II Guards Tank

Belenikhino

Ivanovka

XX
II SS PzGren Div
(*Das Reich*)

Distribution of Soviet Attacking AFVs

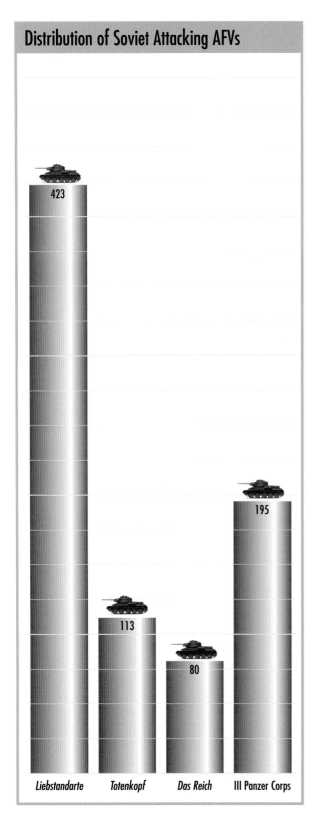

423	113	80	195
Liebstandarte	Totenkopf	Das Reich	III Panzer Corps

hidden in a large orchard on the slopes of Hill 230.5 just south of Prokhorovka. The position gave a clear view of the enemy deployment, and just as the pre-planned 15 minute Soviet artillery and Katyusha bombardment began at 06:00, the German guns opened up and *Leibstandarte*'s armour started its advance. In response to the hail of incoming fire, the panzers fanned out into open order as the 290 tanks of XVIII and XXIX Tank Corps forming the first wave of the Soviet attack started their engines, the clouds of exhaust smoke clearly visible to the German tank crews.

Rotmistrov realized that there was no chance of achieving tactical surprise and that his attack would have to go in quickly, before the panzers overran his tanks in their assembly areas. As the Soviet bombardment ended, he gave the codeword 'Stal' (Steel) to order the assault and the mass of T-34s and T-70s moved off. *Shturmoviks* bombed and strafed the German forces in the few minutes before it became impossible to identify targets as the two tank forces closed with each other.

SOVIET RESISTANCE

The panzer crews were understandably temporarily shocked at the sight of the rapidly approaching mass of almost 300 Soviet tanks firing on the move despite being festooned with infantry. The training and experience of *Leibstandarte*'s tank crews soon kicked in and they halted to select their targets as the Soviet armour came within effective range. Rudolf von Ribbentropp, the son of the Reich Foreign Minister, who commanded a company of Panzer IVs, recalled that: 'As we drove down the forward slope, we spotted our first T-34s, which were apparently trying to outflank us from the left. We halted on the slope and opened fire, hitting several of the enemy. A number of Russian tanks were left burning. For a good gunner 800 metres was the ideal range.'

It was soon apparent that the sheer number of Soviet tanks advancing on narrow frontages was causing its own problems. The very limited room for manoeuvre meant that many vehicles collided as they tried to avoid the burning victims of German fire. (The fate of the infantry carried on the T-34s and T-70s was even worse – those that survived the hits which destroyed their tanks were frequently run down by the milling mass of armour.)

The initial Soviet attacks were finally beaten off by about 09:00, after almost three hours of fierce fighting, but there was only a brief lull before Rotmistrov launched another series of attacks across the entire front. Fierce artillery bombardments covered the advance of more Soviet tanks from Prokhorovka and Jamki. *Leibstandarte*'s divisional history vividly described the action: 'Four of the seven panzers [of 6th Panzer Regiment] were put out of commission at a distance of only about 220 metres. The remaining three panzers joined the ranks of advancing Russian tanks and moved with the pack of them [towards 2nd Panzer Regiment] located about 800 metres to the rear. These three could fire at the Soviets from a distance of 10 to 30 metres and make every shell a direct hit because the Russians could not see through the dust and smoke that there were German tanks rolling along with them. ... There were already 19 Russian tanks standing burning on the battlefield when [2nd Panzer Regiment] opened fire. ... [it] destroyed about 62 T-70s and T-34s in a three-hour long battle that could almost be termed hand-to-hand tank combat.'

At 11:00, a small group of four T-34s took advantage of the confusion to attack *Leibstandarte*'s thinly stretched reconnaissance battalion screening the division's left flank, and broke through. They then charged the divisional artillery positions, knocking out two 150mm (5.9in) guns before they were destroyed by infantry anti-tank teams supported by field artillery firing over open sights. Scarcely had this fighting died away than another series of Soviet attacks were made, at about 11:30. These achieved a local breakthrough near Hill 252 before being repulsed with the usual heavy losses.

Despite the intensity of the morning's fighting, *Leibstandarte* was able to reorganize and resume its advance on Prokhorovka during the afternoon. By this time, XVIII and XXIX Tank Corps had suffered such heavy losses that they were no longer effective combat formations. In desperation, Rotmistrov scraped together a force totalling perhaps 120 T-34s and T-70s drawn from V Guards Mechanized Corps, 10th Guards Mechanized Brigade and 24th Guards Tank Brigade – a scratch force that proved just sufficient to halt the German advance. As the fighting petered out in mutual exhaustion during the evening, the bulk of Fifth Guards Tank Army pulled back to defensive positions in and around Prokhorovka, whilst the majority of *Leibstandarte*'s units found themselves in much the same positions that they had held at the start of the day.

THE EASTERN BATTLEFIELD

The sector to the east of the railway embankment was largely held by *Das Reich,* which was primarily committed to a defensive role, protecting II SS Panzer Corps' long right flank. (III Panzer Corps which should have been acting as flank guard had been badly delayed by strong Soviet field defences and was still 20km/12.4 miles to the south on the morning of the 12th.)

The first Soviet attacks in this sector were made against *Das Reich*'s positions at about 08:30 by elements of II Tank Corps and II Guards Tank Corps totalling perhaps 120 tanks. These were not beaten off until mid-morning, delaying *Das*

A Panzer Mk VI Ausf E waits in a Russian village, somewhere in the Kursk area, July 1943.

Reich's Deutschland Regiment's attack in support of *Leibstandarte*'s assault on the heavily fortified village of Storozhevoe, which was finally taken during the afternoon.

Soviet pressure in this sector was so unremitting that *Deutschland*'s attack was the only significant offensive action that the division was able to undertake throughout 12 July. In a typical attack just after 14:00, 50-plus T-34s advanced from Vinogradovka towards Yasnaya Polyana, but were intercepted by a force of captured T-34s that had been taken into service with *Das Reich*. The scene was witnessed by Sylvester Stadler, commanding the division's *Der Führer* Regiment, who recalled that: 'In a short period of time all 50 tanks … were set ablaze by shells from the captured tanks with German crews. The Soviet tanks each had a barrel of fuel attached to its back. These could be set on fire by a well-aimed shot, and shortly after the

whole tank exploded. Of the Soviet tanks only the command tank … was equipped with a radio. For this reason, that tank was knocked out first. The other crews were perplexed; obviously they did not recognize the T-34s on the hill as their enemy.'

Der Führer itself was subjected to repeated heavy attacks, in one of which two T-34s broke through its defences and advanced into its rear areas where they attacked supply and repair units. One rammed a company field kitchen, where it was destroyed with improvised demolition charges, whilst the other became bogged down and was knocked out by an anti-tank gun rushed to the area from a nearby repair workshop.

As in the other sectors of the battlefield, the fighting gradually died down during the evening as mutual exhaustion set in, with the front line almost where it had been that morning.

'SO ENDS THE BLOODY BUSINESS OF THE DAY' – THE OUTCOME OF THE 12TH

Soviet propaganda was quick to claim that the 12th had been a great victory, issuing wildly inflated claims that the Germans had sustained losses of anything up to 300 AFVs, including 70 Tigers. The truth was embarrassingly different – the daily divisional strength returns prepared for II SS Panzer Corps indicate that a total of 294 AFVs were serviceable at dawn on 12 July; 24 hours later, the corresponding total stood at 251. On the basis of these figures, Rotmistrov's tank crews had knocked out no more than 43 German AFVs, including 11 Tigers – all 10 of *Totenkopf*'s and a single vehicle belonging to *Leibstandarte*. (German 'write-offs' were significantly lower, as all these Tigers and many of the other AFVs were repairable.)

In contrast, the Soviet losses were horrendous – Fifth Guards Tank Army had approximately 850 serviceable AFVs immediately before the battle, but had at best 200 combat-ready tanks and assault guns on 13 July, indicating the loss of at least 650 in one day's fighting. Indeed, German reports of the

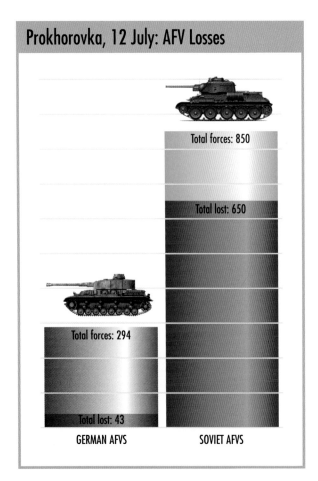

Prokhorovka, 12 July: AFV Losses

Total forces: 850

Total lost: 650

Total forces: 294

Total lost: 43

GERMAN AFVS   SOVIET AFVS

LEFT: Soviet losses for a single day's fighting at Prokhorovka were horrendous, but ultimately sustainable. With so few AFVs available for immediate offensive action, Rotmistrov was forced to go on the defensive until losses were replaced and damaged tanks repaired. German recovery and repair rates were far superior, partly explaining why so many SS AFVs were ready for combat on 13 July.

Prokhorovka: Tank Operations, 12–14 July 1943

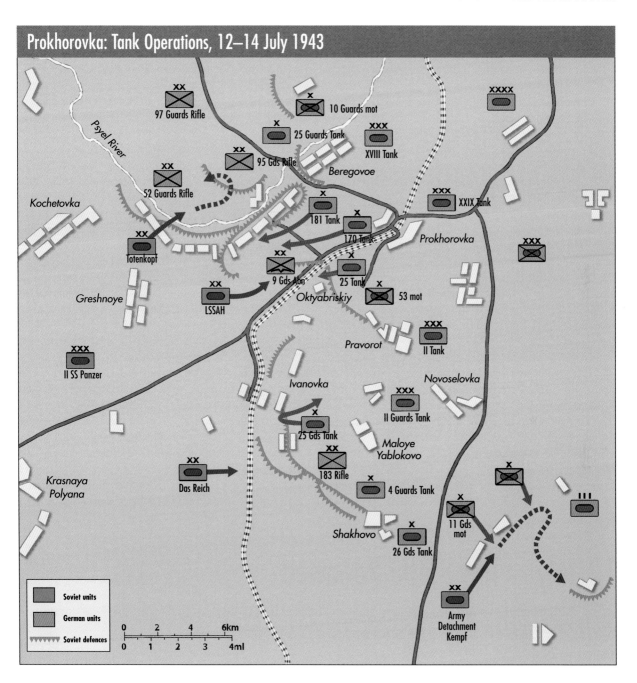

ABOVE: By 11 July, II SS Panzer Corps had achieved a deep penetration of the Soviet defence lines in the south of the Kursk salient. However, the sheer strength of these defences prevented the three *Waffen-SS* divisions from generating sufficient momentum to overcome the resistance of 9th Guards Airborne Division and take Prokhorovka before the arrival of Fifth Guards Tank Army. Even so, Rotmistrov was badly shaken by the extent of the German advance, which turned Fifth Guards Tank Army from the spearhead of a decisive counter-attack into a sacrificial attack force. It was now a question of saving Prokhorovka at all costs; its loss would fatally undermine the Soviet defence of the entire southern half of the Kursk salient. In a series of meeting engagements on 12 July, Fifth Guards Tank Army did indeed halt *Leibstandarte*, but at the price of horrendous losses. Even as it did so, *Totenkopf*'s continued progress northwest of Prokhorovka, combined with Army Detachment *Kempf*'s belated approach from the south, threatened to outflank Rotmistrov's battered forces. *Totenkopf*'s advance was halted only after fierce fighting on 13 July, at which point Hitler closed down the offensive in order to free forces to meet the Allied invasion of Sicily.

massive casualties which they had inflicted seemed so exaggerated that *Obergruppenführer* Paul Hausser commanding II SS Panzer Corps visited *Leibstandarte's* sector of the front to see the situation for himself. His initial scepticism was quickly dispelled when he saw the mass of wrecked Soviet armour, and he counted over 100 knocked-out Soviet tanks, marking each with chalk to keep an accurate tally.

In accordance with standing orders, the senior Soviet commanders phoned Stalin to report on the day's actions. Vasilevsky reported first in his capacity as Stavka representative and had the unenviable task of breaking the news of Fifth Guards Tank Army's losses, which was only slightly mitigated by the fact that II SS Panzer Corps' advance had been halted. When Rotmistrov phoned with a more detailed report, Stalin allegedly rounded on him, shouting 'What have you done to your magnificent tank army?'

Always paranoid, Stalin seems to have become highly suspicious of the competence of the entire senior command structure of the Voronezh Front. Zhukov, his most trusted 'trouble-shooter', was at the HQ of the Bryansk Front, overseeing the preparations for Operation *Kutuzov* (the Soviet offensive to eliminate the Orel salient). Stalin abruptly ordered him to fly out to the Prokhorovka sector and take over from Vasilevsky as the *Stavka* representative. (Vasilevsky was

OPPOSITE: In view of the ferocity of the fighting on 12 July, the Corps' operational AFV strength remained remarkably high. (Though *Leibstandarte* bore the brunt of Fifth Guards Army's attack, the highest losses were sustained by *Totenkopf*, which was advancing through heavily fortified Soviet positions on the Corps' left flank.) The efficient German recovery and repair organization was retrieved most of their knocked-out AFVs, and many were made serviceable within a few days.

transferred to become *Stavka* representative at Southwest Front HQ.) The losses sustained by Fifth Guards Tank Army left Rotmistrov with little option but to go over to the defensive for the next few days. The surviving tanks of the most badly battered armoured units were dug in to support the anti-tank strongpoints, which were hurriedly constructed to protect Prokhorovka. The stronger tank units were held in reserve to counter-attack if a renewed German assault breached the defences.

Rotmistrov had to carefully deploy his remaining strength to counter three main threats:
■ The possibility that *Totenkopf* might continue its drive along the north bank of the Psel and outflank his position.
■ A renewed frontal attack by *Leibstandarte* with the aim of breaking through to Prokhorovka.
■ The belated arrival of III Panzer Corps from the south.

(The 'nightmare scenario' was that *Leibstandarte* might pin his forces in place whilst *Totenkopf* and III Panzer Corps closed in from north and south to carry out a classic encirclement of the entire army.)

13 July – The Battle Renewed

During the night of 12/13 July, von Manstein, Hoth and Hausser agreed that exploitation of *Totenkopf's* advance offered the best chance of unhinging the Soviet position at Prokhorovka.

The division was accordingly ordered to continue its attacks on the 13th with the objective of cutting the Belgorod-Kursk highway north of the Psel. At the same time, *Leibstandarte* would resume its assault on Prokhorovka, whilst *Das Reich* advanced south of the town to make contact with III Panzer Corps.

Rotmistrov had identified *Toptenkopf* as the primary threat and moved 10th Guards Mechanized Brigade to reinforce XXXIII Guards

Rifle Corps, which would bear the brunt of any German attacks in this sector. *Totenkopf* could commit only 54 serviceable panzers to spearhead its renewed offensive because its 20 or so assault guns were still needed to guard the bridges across the Psel. Despite its reduced strength, the division was able to reach its objective by late morning in the face of repeated Soviet attacks. (*Leibstandarte* made limited attacks to coincide with this offensive, which were unable to make much impression on the

II SS Panzer Corps: Operational AFVs, 11–16 July 1943

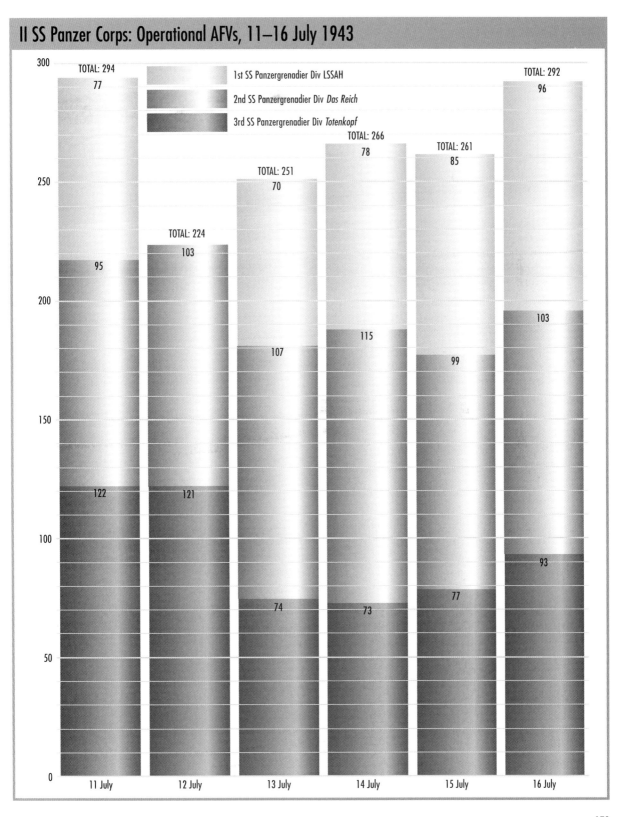

Legend:
- 1st SS Panzergrenadier Div LSSAH
- 2nd SS Panzergrenadier Div *Das Reich*
- 3rd SS Panzergrenadier Div *Totenkopf*

11 July — TOTAL: 294 (77, 95, 122)
12 July — TOTAL: 224 (103, 121)
13 July — TOTAL: 251 (70, 107, 74)
14 July — TOTAL: 266 (78, 115, 73)
15 July — TOTAL: 261 (85, 99, 77)
16 July — TOTAL: 292 (96, 103, 93)

defence positions covering Prokhorovka. These attacks did give limited assistance to *Totenkopf* by forcing the Soviets to divide their fire.) However, constant attacks against the long northern flank of the German salient that had now formed north of the Psel made it impossible to hold the position, and *Totenkopf* was forced to withdraw to a more readily defensible line during the afternoon.

Leibstandarte's main effort was made in the early afternoon, when it launched a two-pronged attack against the XVIII and XXIX Tank Corps with its remaining 50 panzers and 20 assault guns. The principal attack was directed at Soviet positions north-east of the Oktabrisky State Farm, whilst the reconnaissance battalion made a probing attack along the south bank of the Psel near the villages of Andreyevka and Mikhailovka. Both attacks broke down after an hour or so in a maze of newly constructed Russian defences, with minefields causing the greatest damage.

On the southern sector of the front, *Das Reich* was able to make some progress with its attacks along the line Ivanovka–Vinogradovka. It was helped by the defenders having to divert units to face the advance of III Panzer Corps, which was now only 12km (7.45 miles) away from the southern outskirts of Prokhorovka.

13 JULY – WINNERS & LOSERS

The day's actions had exhausted *Totenkopf* and *Leibstandarte* to such a degree that they were temporarily incapable of doing anything more than defending their positions. However, *Das Reich* and III Panzer Corps retained a significant offensive capability and von Manstein believed that there was still a realistic prospect of at least destroying Fifth Guards Tank Army and the remaining Soviet armoured forces in this sector. Zhukov, Vatutin and Rotmistrov were far from convinced that they had weathered the storm. The day's fighting had inflicted further heavy losses on their formations: XVIII Tank Corps had lost 30 per cent of its strength, whilst XXIX Tank Corps losses now stood at a crippling 60 per cent.

However, the key decisions were being made far away from the battlefield in East Prussia at Hitler's Rastenburg HQ. The Allied invasion of Sicily had begun on 9/10 July with very little effective resistance from the Italian forces on the island. Hitler was convinced that Mussolini's regime was on the point of being overthrown by a military coup, after which Italy would make a separate peace with the Allies or possibly even change sides. On 13 July, von Manstein and von Kluge, the commanders of Army Groups South and Centre arrived at Rastenburg for a conference with Hitler, who announced that the critical situation in Italy compelled him to call off Operation 'Citadel' so that II SS Panzer Corps could be rushed to Italy to bolster Mussolini's position.

Whilst von Kluge was relieved at the decision, which would free his forces to counter the Red Army's attacks on the Orel salient, von Manstein was appalled at what he saw as the premature abandonment of the chance to decisively defeat the First Tank Army and Fifth Guards Tank Army around Prokhorovka, which he believed were the entire Soviet armoured reserves. He emphasized

BELOW: As the tables show, the defensive part of the Kursk battle was much less costly in casualties and equipment for the Red Army than the consequent offensives at Orel and Belgorod–Kharkov, with the Soviets losing 177,847 men between 5–23 July, but another 1.8 million in operations up to the end of September 1943.

Soviet Equipment Losses, Summer 1943

OPERATION	DATE	SMALL ARMS WEAPONS	TANKS AND ASSAULT GUNS	GUNS AND MORTARS
Kursk defence	5–23 July	70,800	1614	3929
Orel offensive	12 July–18 August	60,500	2586	892
Belgorod–Kharkov offensive	3–23 August	21,700	1864	423
Smolensk offensive	7 August–2 October	33,700	863	234
Donbass offensive	13 August–22 September	37,900	886	814
Chernigov–Poltava	26 August–30 September	48,000	1140	916

Fourth Panzer Army's successes since 5 July – it had taken 24,000 prisoners and virtually annihilated up to 10 tank and mechanized corps, with the destruction or capture of 1800 AFVs, 267 field guns and 1080 anti-tank guns.

Despite his concerns about the looming crisis in Italy, Hitler was sufficiently impressed by von Manstein's arguments to authorize a temporary continuation of the attacks around Prokhorovka, whilst closing down Ninth Army's offensive against the northern face of the Kursk salient. (However, very shortly afterwards he robbed the operation of any real chance of success by overruling von Manstein and ordering XXIV Panzer Corps to defend the Donbass against an anticipated Soviet offensive instead of reinforcing III Panzer Corps for its renewed attack on Fifth Guards Tank Army.)

14–17 July – Operation 'Roland'

Although bitterly disappointed at the absence of XXIV Panzer Corps, von Manstein believed that there was still a good chance of mauling Rotmistrov's forces. He and Hoth agreed that *Totenkopf* and *Leibstandarte* should remain on the defensive, whilst *Das Reich* and III Panzer Corps concentrated their attacks south of Prokhorovka, encircling the Soviet forces deployed between the Lipovyi Donets and the North Donets.

The first stage of the operation began at 04:00 on 14 July with an attack by *Das Reich* against the village of Belenikhino, which fell to the *Der Führer* Regiment after air attacks followed by fierce hand-to-hand fighting. The division's panzer regiment then led an assault on the village of Ivanovka, which was taken that afternoon and inflicted further heavy losses on the already weakened II Guards Tank Corps, which was forced to withdraw eastwards.

At the same time, III Panzer Corps was attacking northwards in an attempt to break through the network of strongpoints set up by the newly arrived 31st and 32nd Anti-Tank Brigades and link up with *Das Reich*. These attacks were costly and frustrating; although the defenders were under orders to fall back steadily to avoid encirclement, they lost substantial numbers of anti-tank guns, which could not be withdrawn in time.

The following day, *Das Reich* resumed its advance in driving rain against fierce opposition, but was heartened by a report that III Panzer Corps was now making real progress and the two formations finally made contact early that afternoon. Despite this success, *Das Reich* was unable to break through the maze of minefields, anti-tank ditches and strongpoints defending the town of Pravorot.

Throughout this period, there was no respite for Fifth Guards Tank Army, which Zhukov ordered to maintain pressure on *Totenkopf* and *Leibstandarte*, to prevent any advance or redeployment that might yet jeopardize the defence of Prokhorovka. Between 14 and 17 July, both formations were subjected to repeated artillery and air bombardments. Soviet tank losses had been so heavy that only small scale company or battalion strength probing attacks could be made, all of which were beaten off.

The German failure to take Pravorot on the 15th finally convinced von Manstein and Hoth that there was no longer any realistic chance of Operation 'Roland' succeeding, and on 17 July the order was given to prepare for a phased withdrawal to the positions held before the start of Operation 'Citadel'.

DAS REICH'S LOSSES

Throughout the offensive, the number of *Das Reich*'s operational AFVs varied wildly, but surprisingly few damaged vehicles were complete write-offs. (During the period 5–17 July, the total of such losses sustained by the division amounted to only one Panzer III, six Panzer IVs, a single Tiger and one StuG III.) This was largely due to the highly efficient divisional recovery and workshop teams, who were helped by the fact that for much of this time II SS Panzer Corps was either advancing or successfully defending its positions. This made the recovery of damaged vehicles relatively straightforward – it was far harder to retrieve and repair knocked-out tanks during a retreat. German 'total losses' of AFVs rose sharply during the period after Kursk when they were in almost constant retreat, and it was impossible to recover many repairable vehicles, which had to be blown up to prevent capture.

WHO WON?

For many years after the war, Soviet propaganda was accepted by most historians. It certainly generated some memorable phrases, such as 'the Tigers are burning' and 'the death ride of the panzers', but these had little basis in fact. Each side's losses for 12 July have already been given, but those for the entire period are even more revealing. Documents held by the office of the Inspector General of Panzer Troops indicate that just 33 tanks and assault guns of II SS Panzer Corps were totally destroyed during the period 5–17 July. Comparable figures for Fifth Guards Tank Army are far higher – on 17 July 1943 Major General Baskakov, the Army's Chief of Staff, reported that its 'irretrievable losses' for the period 12–16 July totalled 334 tanks and assault guns, approximately 40 per cent of its AFVs.

II SS Panzer Corps certainly demonstrated immense professionalism and expertise in inflicting 11 times its own tank losses whilst attacking a powerful enemy force that benefited from elaborate field defences, but the crucial point was that the Red Army could afford such losses. In July and August 1943, Soviet factories completed roughly 4000 tanks, more than sufficient to replace the 1600 or so which had been lost during the entire Kursk Offensive. Significantly, the Red Army was able to deploy a total of 2832 AFVs in early August for the offensive which retook Belgorod and Kharkov (Operation *Polkovodets Rumyantsev*), giving it a 5:1 superiority over the defending German forces.

OPPOSITE: On 4 July 1943, *Das Reich's* 'ration strength' was given as 20,659, whilst its combat strength (personnel in its panzer, panzergrenadier, anti-tank and engineer units) was only 7350. Whilst there are no specific figures for the rest of II SS Panzer Corps, their figures are unlikely to have been very different. In all three divisions, most of the casualties shown here were sustained by the 'combat arms', which suffered far higher percentage losses than those for the divisions as a whole.

BELOW: The majority of these figures are taken from the division's reports which show details of operational AFVs as at the early evening of each day. The numbers of operational vehicles were generally at their lowest at this time because damaged tanks and StuGs would be recovered during the evening and worked on overnight, so that units could field the maximum possible number of vehicles next morning.

Das Reich: Operational AFVs 4–18 July

DATE	COMMAND TANKS	PANZER III	PANZER IV	TIGER	STUG III	T-34
4 July	8	48	30	12	33	18
5 July	8	53	27	11	21	16
7 July	6	48	16	7	14	15
8 July	7	43	25	6	7	14
9 July	7	31	13	1	26	7
10 July	7	33	15	1	26	7
11 July	7	34	18	1	27	8
12 July	6	42	18	2	27	8
13 July	8	43	20	1	24	11
14 July	8	41	25	4	25	12
15 July	7	37	17	2	23	13
16 July	7	37	18	5	25	11
17 July	6	36	24	9	25	17
18 July	7	36	24	9	28	17

II SS Panzer Corps Manpower Losses, 11–16 July 1943

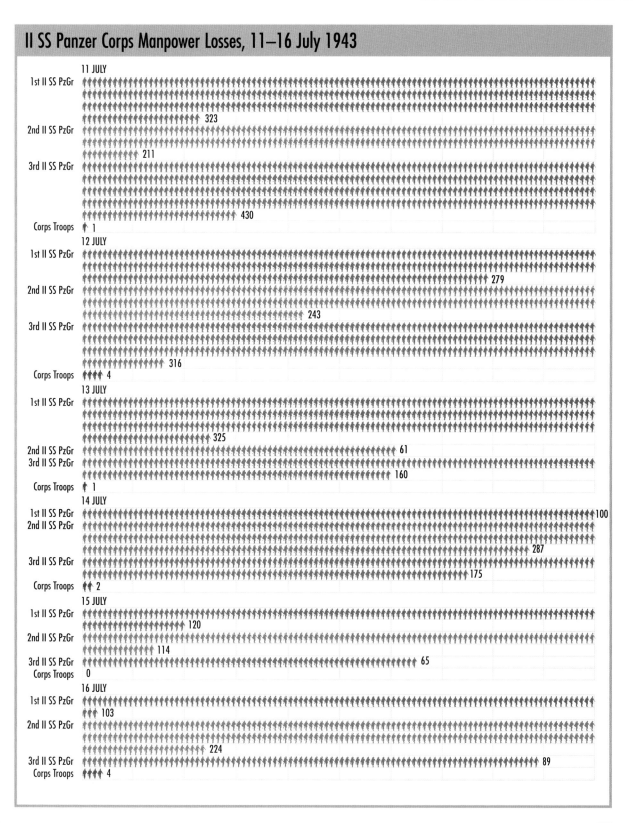

11 JULY

1st II SS PzGr

2nd II SS PzGr — 323

3rd II SS PzGr — 211

Corps Troops — 430 / 1

12 JULY

1st II SS PzGr

2nd II SS PzGr — 279

3rd II SS PzGr — 243

Corps Troops — 316 / 4

13 JULY

1st II SS PzGr

2nd II SS PzGr — 325

3rd II SS PzGr — 61

Corps Troops — 160 / 1

14 JULY

1st II SS PzGr

2nd II SS PzGr — 100

3rd II SS PzGr — 287

Corps Troops — 175 / 2

15 JULY

1st II SS PzGr — 120

2nd II SS PzGr — 114

3rd II SS PzGr — 65

Corps Troops — 0

16 JULY

1st II SS PzGr — 103

2nd II SS PzGr — 224

3rd II SS PzGr — 89

Corps Troops — 4

Chapter 5
Aftermath: 1943

Soviet forces in the north of the Kursk salient recovered quickly
from the German attack and launched their own offensive, Operation *Kutuzov*,
against Orel on 12 July.

Orel was strongly defended – falling on 3/4 August only after
Third Guards Tank Army and Fourth Tank Army had been committed to the assault.
Whilst this prolonged battle was under way, the Red Army's Southern Front began an
offensive on 17 July against the German defences along the River Mius protecting
Army Group South's extreme southern flank. By the following day, Soviet forces
spearheaded by Fifth Shock Army had achieved a breakthrough, despite
counter-attacks by 16th Panzergrenadier Division. The Soviet momentum was
maintained until 22 July and created a dangerous bridgehead on the west bank
of the Mius despite further counter-attacks by 23rd Panzer Division.

OPPOSITE: Neatly camouflaged to blend in with a building, an SS Tiger Ausf E waits for the order to move forward following
a Soviet air attack. The commander sitting in the turret is wearing summer pattern camouflage, which was permissible on the
Eastern Front during the summer months.

Operation *Polkovodets Rumyantsev*

Das Reich and *Totenkopf* had received orders to prepare for redeployment to the south on 17 July, but did not arrive at the Mius front (in company with 3rd Panzer Division) until 27 July.

After a brief spell of reconnaissance and preparation, these divisions counter-attacked on 30 July against what was now a well-fortified Soviet bridgehead. After much hard fighting, *Das Reich* finally achieved a decisive breakthrough on 1 August, which it was then able to exploit in company with 16th Panzergrenadier Division and 23rd Panzer Division, splitting the bridgehead in two. On the following day, the Red Army began a withdrawal to the east bank, although the remnants of the bridgehead were not finally eliminated until 10 August.

Stavka had intended that Operation *Polkovodets Rumyantsev*, the Soviet attack on Belgorod and

Kharkov, should begin as soon as the German offensive against the southern sector of the Kursk salient had been halted. However, the battering that *Das Reich* and the rest of Hoth's Fourth Panzer Army had inflicted on Fifth Guards Tank Army and other Soviet formations meant that this offensive could not begin until 3 August.

Belgorod fell on 5 August, but Kharkov was more strongly defended, with its garrison reinforced by *Das Reich* with 96 Panthers, 32 Tigers and 25 assault guns. When Rotmistrov's newly re-equipped Fifth Guards Tank Army attacked on 17 August in an attempt to encircle the city, its assaults were clumsily delivered without proper air, artillery or infantry support against well-prepared defences held by XI Infantry Corps and *Das Reich*. With excellent *Luftwaffe* support, the defenders were able to give a 'textbook' demonstration of all-arms co-operation, in contrast to the poorly co-ordinated Soviet attacks.

Scattered across a cereal field, vehicles from an SS panzergrenadier unit wait for the order to move out. The tarpaulin provides a sunshade and is in turn draped with a large swastika flag for the purpose of air recognition. The SdKfz 250/1 pictured here appears to be the troop carrying version.

General Raus commanding XI Infantry Corps described how 'the Russian tanks had been recognized while they were still assembling in the villages and flood plains of a brook valley. Within a few minutes heavily laden Stukas came on in wedge formation and unloaded their cargoes of destruction in well timed dives on the enemy tanks caught in this congested area. Dark fountains of earth erupted skyward and were followed by heavy thunderclaps and shocks that resembled an earthquake. These were the heaviest, two-ton bombs, designed for use against battleships, which were all that *Luftflotte* 4 had left to counter the Russian attack. Soon all the villages occupied by Soviet tanks lay in flames. A sea of dust and smoke clouds illuminated by the setting sun hung over the brook valley, while dark mushrooms of smoke from burning tanks stood out in stark contrast. This gruesome picture bore witness to an undertaking that left death and destruction in its wake, hitting the Russians so hard that they could no longer launch their projected attack that day, regardless of Stalin's order. Such a severe blow inflicted on the Soviets had bought badly needed time for XI Infantry Corps to reorganize.

'On 20 August the Soviets avoided mass groupings of tanks, crossed the brook valley simultaneously in a number of places, and disappeared into the broad cornfields that were located ahead of our lines, ending at the east-west rollbahn several hundred metres in front of our main battle line. Throughout the morning Soviet tanks worked their way forward in the hollows up to the southern edges of the cornfields, then made a mass dash across the road in full sight. *Das Reich*'s Panthers caught the leading waves of T-34s with fierce defensive fire before they could reach our main battle line. Yet wave after wave followed, until Russian tanks flowed across in the protecting hollows and pushed forward into our battle positions. Here a net of anti-tank and flak guns, Hornet 88mm tank destroyers, and Wasp self-propelled 105mm field howitzers trapped the T-34's, split them into small groups, and put large numbers out of action. The final waves were still attempting to force a breakthrough in concentrated masses when the Tigers and StuG III self-propelled assault guns, which represented our mobile reserves behind the front, attacked the Russian armour and

repulsed it with heavy losses. The price paid by the Fifth Guards Tank Army for this mass assault amounted to 184 knocked out T-34s.'

NIGHT ATTACKS

Rotmistrov then attempted night attacks on two consecutive nights, again without success, bringing his total tank losses to 420 in five days of combat. It was only on 22 August that the defenders withdrew from Kharkov to avoid being cut off. As the withdrawal got underway, *Das Reich*'s Panther battalion was ordered to counter-attack a particularly dangerous breakthrough by the 110 T-34s of 24th Guards Tank Brigade southwest of Kharkov. It destroyed 53 of them in a three-hour action.

The Germans were now being steadily forced westwards by growing Soviet pressure. Von Manstein had argued strongly for a timely, phased withdrawal to the Dnieper, which would shorten Army Group South's line from almost 965km (600 miles) to 725km (450 miles), but Hitler repeatedly refused to authorize such a retreat until mid-September, when his hand was forced by a series of Soviet breakthroughs. *Das Reich* acted as the rearguard for the Eighth Army in its retreat to the Dnieper via Poltava, the site of the only bridge across the River Vorskla. On 20 September, No.3 Battery of the division's StuG battalion commanded by *Obersturmführer* Wolfgang Röhder fought a fierce action to defend the bridge against repeated Soviet attacks whilst it was prepared for demolition. He held the far bank for several hours, withdrawing at the last possible moment, only a matter of minutes before the bridge was blown as Russian infantry began to cross. (In December 1943, Röhder was awarded the Knight's Cross for his leadership in this action.)

DAS REICH COUNTER-ATTACKS

Das Reich was able to withdraw across the Dnieper, but was almost immediately thrown into action in an attempt to eliminate Soviet bridgeheads on the west bank. The first such bridgehead was established on 22 September 1943 at the confluence of Dnieper and Pripyat rivers, on the northern part of the front. On 24 September, another bridgehead was created near Dneprodzerzhinsk, another on 25 September near

Das Reich: Complement, 30 August 1943

EQUIPMENT	STRENGTH
Rations strength	13,592
Combat strength	5692
Combat-ready tanks	22
Tanks in workshop	138
Assault guns combat ready	22
Assault guns in workshop	6
10.5cm howitzers	24
Wespe	12
15cm howitzers	8
Hummel	4
10cm gun	4
15cm NbW	0
2cm Flak	37
3.7cm Flak	6
8.8com Flak	12
Med PaK	34
Heavy PaK	28
7.5cm LeIG	25
15cm sIG	12
8cm GrW	80
Heavy MG	103
Light MG	1388
Rifles and MPs	8685

Dnepropetrovsk and yet another one on 28 September near Kremenchug. By the end of the month, 23 bridgeheads had been formed on the west bank, some of them 10km (6.2 miles) wide and 1–2km (0.6–1.2 miles) deep. All these attracted fierce German counter-attacks, but managed to hold out with massive fire support from Soviet artillery on the east bank of the river.

One of the most experienced German commanders, General von Mellenthin, commented that: 'Bridgeheads in the hands of the Russians are grave danger indeed. It is quite wrong not to worry about bridgeheads, and to postpone their elimination. Red Army bridgeheads, however small and harmless they may appear, are bound to grow into formidable danger-points in a very brief time and soon become insuperable strong points. A Soviet bridgehead, occupied by a company in the evening, is sure to be occupied by at least a regiment the following morning and during the night will become a formidable fortress, well-equipped with heavy weapons and everything necessary to make it almost impregnable. No artillery fire, however violent and well concentrated, will wipe out a Russian bridgehead which has grown overnight. This Russian principle of "bridgeheads everywhere" constitutes a most serious danger and cannot be overrated.'

Das Reich veterans would ruefully agree with his assessment. The division was being steadily worn down, with the panzergrenadiers taking the brunt of the losses – to the extent that many of their companies were reduced to platoon strength. By a stupendous effort, the division eliminated particularly dangerous Soviet bridgeheads near Grebeni and Juschki in ferocious fighting between 28 September and 5 October. Despite this achievement, there were insufficient German forces of Das Reich's calibre to deal with the rash of Soviet bridgeheads and by mid-October, the forces assembled in these areas were strong enough to go launch their own offensives coupled with diversionary attacks in the south to draw German forces away from Kiev. At the end of the operation, the Red Army controlled a bridgehead 300km (186 miles) wide and up to 80km (50 miles) deep, whilst in the far south, Army Group A was now cut off in the Crimea.

AUTUMN DELAYS

German tactical expertise could still impose serious delays on the Soviet advance – on 1 November, Das Reich destroyed its 2000th tank. Its efforts had contributed to ensuring that for much of October 1943 Russian forces were penned into the 'Bukrin Bend' of the Dnieper. However, the Red Army was becoming far more skilful and was able to secretly redeploy these forces northwards to the tiny bridgehead across the Dneiper at Lyutlezh, just

OPPOSITE: Work on the Dnieper defence line, which formed part of the Panther–Wotan Line or the Eastern Wall, had begun as early as 11 August 1943. Although in theory fortifications were to be erected along the length of the Dnieper River, resources were totally inadequate for such a massive project, and defence works were concentrated in sectors where Soviet assault crossings were most likely to be attempted. On 15 September 1943, Hitler finally authorized Army Group South to fall back to the Dnieper defence line and a deadly race ensued as the Red Army attempted to cut off the German line of retreat to the river. Despite the odds, von Manstein managed to get the bulk of Army Group South safely across the Dnieper.

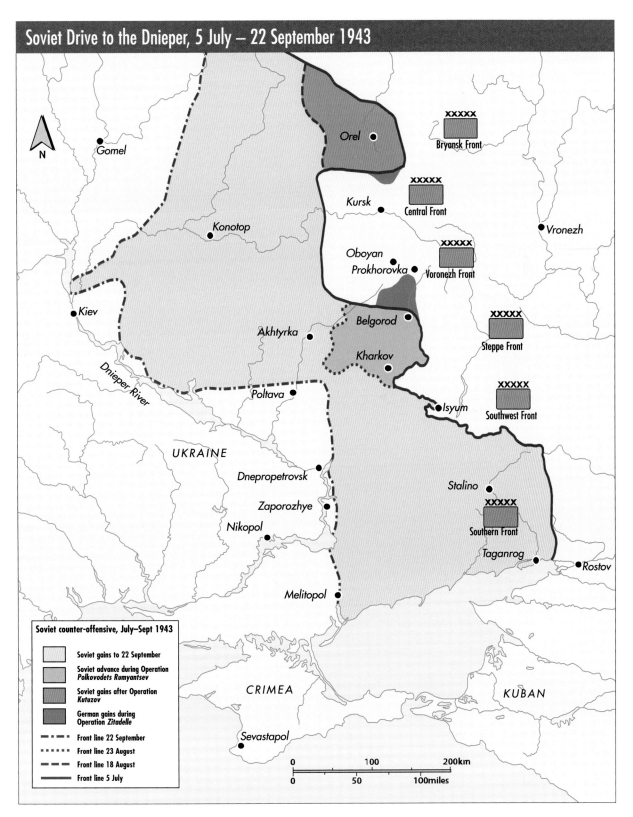

Soviet Drive to the Dnieper, 5 July – 22 September 1943

N

Gomel

Orel

XXXXX
Bryansk Front

Kursk

XXXXX
Central Front

Vronezh

Konotop

Oboyan
Prokhorovka

XXXXX
Voronezh Front

Kiev

Akhtyrka

Belgorod

XXXXX
Steppe Front

Dnieper River

Kharkov

Poltava

Isyum

XXXXX
Southwest Front

UKRAINE

Dnepropetrovsk

Stalino

Zaporozhye

XXXXX
Southern Front

Nikopol

Taganrog

Rostov

Melitopol

Soviet counter-offensive, July–Sept 1943

Soviet gains to 22 September

Soviet advance during Operation *Polkovodets Rumyantsev*

Soviet gains after Operation *Kutuzov*

German gains during Operation *Zitadelle*

— · — · — Front line 22 September

· · · · · · · · Front line 23 August

– – – – Front line 18 August

——— Front line 5 July

CRIMEA

KUBAN

Sevastapol

0   100   200km

0   50   100miles

Soviet Advance to Dnieper

Soviet Advance to Dnieper
5 July – 1 December 1943

- - - - - - Soviet front line, 5 July
⟵ Soviet movements to 1 September
———— Soviet front line, 1 September
⟵ Soviet movements to 1 October
········· Soviet front line, 1 October
⟵ Soviet movements to 1 December
———— Soviet front line, 1 December
⟶ German counter-attacks
⊓⊔⊓⊔ Wotan defensive line

upstream of Kiev. By 3 November, the move was complete and 7th Artillery Breakthrough Corps unleashed a bombardment by 2000 guns, mortars and rocket launchers before Thirty-Eighth Army went in to the attack to open the way for General Rybalko's Third Guards Tank Army. Rybalko's forces maintained the advance throughout the night, with the tanks using their headlights and sirens to dazzle and demoralise the defenders. Despite German counter-attacks, the momentum of the offensive was sustained and V Guards Tank Corps led a successful assault on Kiev on the night of 5th/6th November.

The Fourth Panzer Army attempted to halt the Soviet advance, but Vatutin's newly re-designated 1st Ukrainian Front took Zhitomir and Korosten, cutting the rail link between Army Groups Centre and South. At this critical point von Manstein counter-attacked with XXXXVIII Panzer Corps (1st, 7th, 19th and 25th Panzer Divisions, plus *Leibstandarte* and elements of *Das Reich*).

LEFT: 5 JULY – 1 DECEMBER 1943

After the recapture of Kiev, the Red Army was content to clear the rest of the Dnieper in the south, and recapture a few significant places to the north. Now the ground was frozen, the Russian winter again chilled the hearts of the German forces. The Steppe and Southwestern Fronts drove across the river and formed a wide and deep penetration pointing at Krivoi Rog and Kirovograd. Southern Front reached the mouth of the Dnieper and effectively shut off all German forces left in the Crimea. To the north, Generals Vatutin and Rokossovsky had driven their fronts as far as Korosten and the eastern edge of the Pripet Marshes, and Sokolovsky had taken – at great cost – the massive defensive bastion that the Germans had made of Smolensk.

This force recaptured Zhitomir, fighting a fierce battle with VII Guards Tank Corps before the deep mud created by the autumn rains temporarily halted operations. Both sides took advantage of the enforced lull to refit, but XXXXVIII Panzer Corps took the initiative. As soon as the ground froze in early December, it launched an attack north of Zhitomir with the aim of encircling Sixtieth Army, which hurriedly withdrew from Korosten. The situation was so critical that Stavka transferred First Tank Army and Eighteenth Army to First Ukrainian Front. These reinforcements allowed Vatutin to halt the German attack and return to the offensive – by mid-December, it seemed that both sides were exhausted and XXXXVIII Panzer Corps was withdrawn to rest and refit. However, Vatutin was determined to exploit his numerical superiority and renewed his attacks on 24 December – these made good progress, and as the year ended, his forward units were approaching the 1939 Polish frontier.

By this time, *Das Reich* had been 'bled white' – on 12 December the very last combat-capable personnel were sent forward by its field replacement battalion, and unit commanders were using HQ staff to plug the worst gaps in its fragile front line. This was, however the turning point – it had been officially recognized that the division had to be refitted and orders were issued for its withdrawal to East Prussia to reform as 2nd SS Panzer Division *Das Reich*. Its sector of the front was handed over to *Leibstandarte* and all its remaining combat elements were reformed as the *Kampfgruppe* (battle group) *Das Reich*, whilst the remainder of the division began the long journey back to Germany. (The *Kampfgruppe* remained in action on the Eastern Front until April 1944, when it was withdrawn to rejoin the division.)

EPILOGUE

Das Reich was to fight on, seeing action in the Normandy campaign, the Ardennes offensive and at Lake Balaton in Hungary before surrendering to US forces in Austria at the end of the war.

A company of Panzer IV Ausf H tanks moves up to its start line. The camouflage pattern is dark olive green sprayed over dark yellow to blend in with the summer colours of the steppe. The turret and chassis skirt armour carry the tactical number and Balkan cross in white.

Bibliography

BOOKS

Bergstrom, Christer. *Kursk, the Air Battle: July 1943*. Ian Allan Publishing Ltd, 2007.

Carell, Paul. *Scorched Earth, Hitler's War on Russia, Volume 2*. George G. Harrap & Co. Ltd, 1970.

Forczyk, Robert. *Panther versus T-34: Ukraine 1943*. Osprey Publishing, 2007.

Glantz, David M. & Orenstein, Harold S. *The Battle for Kursk 1943: The Soviet General Staff Study*. Frank Cass Publishers, 2002.

Healy, Mark. *Zitadelle, the German Offensive Against the Kursk Salient, 4–17 July 1943*. Spellmount, The History Press Ltd, 2010.

Kershaw, Robert. *Tank Men: The Human Story of Tanks at War*. Hodder & Stoughton Ltd, 2009.

Nafziger, George F. *The German Order Of Battle: Waffen SS And Other Units In World War II*. Da Capo Press, 2000.

Rottman, Gordon L. *German Pionier 1939–45: Combat Engineer of the Wehrmacht*. Osprey Publishing, 2010.

Schneider, Wolfgang. *Panzer Tactics. German Small-Unit Armor Tactics in World War II*. Stackpole Books, 2005.

Vanags-Baginskis, Alex and Watanabe, Rikyu. *Aggressors Volume 1: Tank Buster Vs. Combat Vehicle*. Airlife Publishing Ltd, 1990.

Weidinger, Otto. *Das Reich IV, 1943*. J. J. Fedorowicz, 2008.

Williamson, Gordon. *Panzer Crewman 1939–45*. Osprey Publishing, 2002.

Winchester, Charles D. *Hitler's War on Russia*. Osprey Publishing, 2007.

Zetterling, Niklas & Frankson, Anders. *Kursk 1943: A Statistical Analysis*. Frank Cass Publishers, 2000.

WEBSITES

Das Reich 2nd SS Panzer Division:
http://www.dasreich.ca/index.html
A wide-ranging website covering many aspects of the division's history.

Engines of the Wehrmacht:
http://www.german.o5m6.de/
Oliver Missing's excellent website covers a steadily growing range of German trucks and miscellaneous vehicles.

Kursk – July 1943
http://dspace.dial.pipex.com/town/avenue/vy75/kursk.htm

The Kursk Battle
http://rkkaww2.armchairgeneral.com/battles/kursk43.htm#Oleinikov
A part of the excellent RKKA in World War II website.

Lone Sentry:
http://www.lonesentry.com/
Contains an invaluable collection of wartime Allied intelligence publications covering a vast range of topics related to Axis equipment and tactics.

The Tiger I Information Center:
http://www.alanhamby.com/tiger.html
One of the most detailed websites on all aspects of the Tiger I.

Guide to Symbols

UNITS

XXXX	ARMY
XXX	CORPS
XX	DIVISION
X	BRIGADE
I I I	REGIMENT
I I	BATTALION
I	COMPANY
•••	PLATOON
••	SECTION
•	SQUAD/TEAM

TYPE

CAVALRY	
INFANTRY	
MOTORIZED INFANTRY	
RECONNAISSANCE	
ARMOUR/TANK	
SELF-PROPELLED ARTILLERY	
ARTILLERY	
ANTI-TANK	
AIR DEFENCE	
MORTAR	

MEDICAL	
ENGINEERS	
SIGNALS	
TRANSPORT	
SUPPLY	
MAINTENANCE	

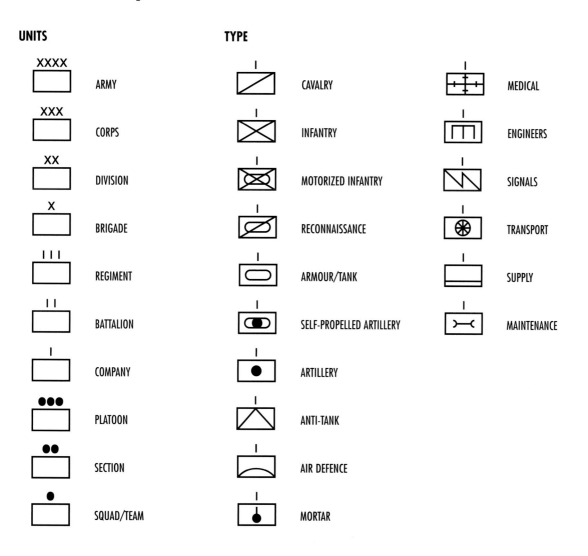

DAS REICH DIVISIONAL SYMBOLS

The *Wolfsangel* (Wolf's Hook) is a germanic symbol dating back to the Middle Ages, thought in folklore to represent a kind of iron trap used to catch wolves. A horizontal version of the symbol as used by *Das Reich* was known as 'the werewolf'. It was often painted on to vehicles without a surrounding shield.

As the 2nd SS Division, at Kursk *Das Reich* was given a symbol with two vertical uprights, compared to the single upright of *Leibstandarte* and three vertical strokes of *Totenkopf*.

Index

Page numbers in *italics* refer to illustrations.

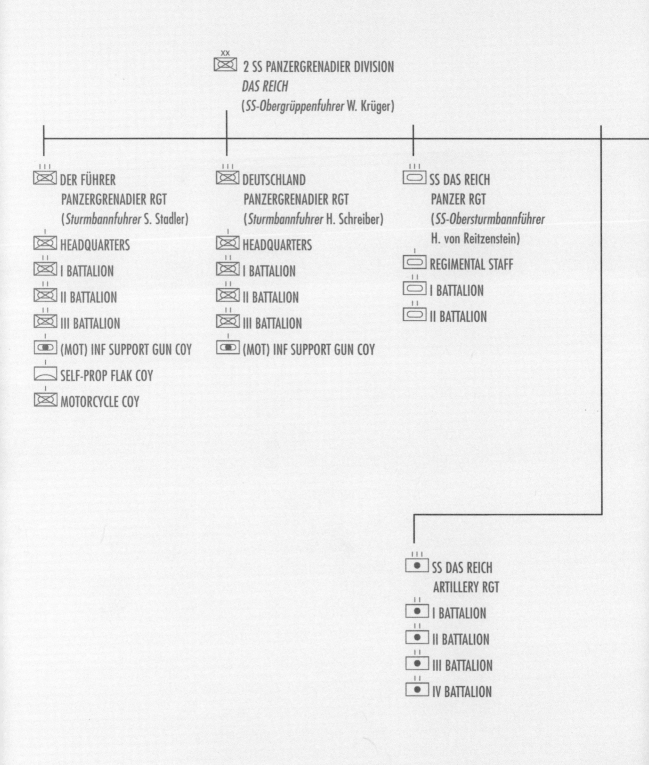

XX
⊠ 2 SS PANZERGRENADIER DIVISION
DAS REICH
(*SS-Obergrüppenfuhrer* W. Krüger)

⊠ DER FÜHRER
PANZERGRENADIER RGT
(*Sturmbannfuhrer* S. Stadler)
⊠ HEADQUARTERS
⊠ I BATTALION
⊠ II BATTALION
⊠ III BATTALION
⊡ (MOT) INF SUPPORT GUN COY
◠ SELF-PROP FLAK COY
⊠ MOTORCYCLE COY

⊠ DEUTSCHLAND
PANZERGRENADIER RGT
(*Sturmbannfuhrer* H. Schreiber)
⊠ HEADQUARTERS
⊠ I BATTALION
⊠ II BATTALION
⊠ III BATTALION
⊡ (MOT) INF SUPPORT GUN COY

▭ SS DAS REICH
PANZER RGT
(*SS-Obersturmbannführer*
H. von Reitzenstein)
▭ REGIMENTAL STAFF
▭ I BATTALION
▭ II BATTALION

⊡ SS DAS REICH
ARTILLERY RGT
⊡ I BATTALION
⊡ II BATTALION
⊡ III BATTALION
⊡ IV BATTALION